INSIDE THE CLOSED WORLD OF THE BRAIN

HOW BRAIN CELLS CONNECT, SHARE AND DISENGAGE—AND WHY THIS HOLDS THE KEY TO ALZHEIMER'S DISEASE

MARGARET T. REECE PhD

REECE BIOMEDICAL CONSULTING LLC
MANLIUS, NEW YORK

Margaret T. Reece PhD/Reece Biochemical Consulting LLC
8195 Cazenovia Road
Manlius, New York 13104
www.medicalsciencenavigator.com

Book Layout ©2013 BookDesignTemplates.com; Cover Image ©Viktoriya, Shutterstock.com

Ordering Information:
Quantity sales. Special discounts are available on quantity purchases by corporations, associations, and others. For details, contact the "Special Sales Department" at the address above.

Inside the Closed World of the Brain/ Margaret T. Reece PhD — 1st ed.
ISBN 978-0-9963513-0-0

For all who are dedicated to eradication of the long goodbye that is Alzheimer's disease.

Preface

MOST EVERYONE HAS HEARD of Alzheimer's disease, but few know much about it. Because I teach human physiology, friends, acquaintances, students, family members and strangers frequently ask me questions about it. What is Alzheimer's disease? How is Alzheimer's disease different than just getting old? Can I avoid Alzheimer's disease by keeping my cholesterol level under control? Coming up with a clear accurate answer to these and similar questions over coffee or lunch is a challenge. First, words that describe how a brain routinely works require explanation. Second, some myths about the human brain must be dispelled. Third, the phases of Alzheimer's disease prior to the appearance of symptoms need to be described. My goal with this book is to provide readers with state-of-the-art knowledge of how brain cells normally work together and where they may go astray to establish Alzheimer's disease. There is considerable reason to believe that ongoing research efforts will produce ways to prevent, or sufficiently slow, Alzheimer's disease so that people in the future can live a normal lifespan without experiencing this form of dementia.

Margaret T. Reece, PhD

Introduction

THE TEMPTATION TO READ chapter 8, *"When It All Goes Wrong—Alzheimer's Dementia"* first is understandable. For readers with a background in neuroscience, that approach should not be a problem. Others will find reference throughout chapter 8 to earlier chapters with needed background material. Chapters 1-7 are organized to progressively build a basic vocabulary for newcomers to the science. Medical students will find numerous facts on every page that are extracted from actual Step 1 exam questions.

Chapter 1 presents tactics for quickly learning the necessary words. The second chapter provides an explanation of the general organization of the human brain both at the visual and microscopic level. The next chapter describes the brain's elaborate system for quality control of the fluids surrounding its cells. Two chapters are devoted to neurons, the superstars of the brain cell community. The first discusses where neurons get their electricity and the second explains how neurons communicate with each other. In chapter 6 the brain's other, non-neuron, cells are introduced, and their partnership with neurons is explained. In chapter 7, the consensus within psychology and neuroscience is presented concerning critical elements of memory formation and language acquisition.

Glossary and Further Reading sections are included at the end. Further Reading is a partial list of the original papers consulted in creating this book.

CONTENTS

"Some things need to be believed to be seen."

STEVE JOBS

[1]

Tips & Tricks for Learning Scientific Language

THE STRANGE WORDS USED in anatomy and physiology make it difficult to follow discussions of the science. Because scientific language is an obstacle for many, this book begins by describing the secret to understanding the words needed to learn about what happens inside the human brain.

Human anatomic names were assigned when scholars wrote and lectured in Classical Latin. Classical Latin was the universal language of large segments of the western scientific world from the time of the Roman Empire (*Figure 1-1*) through the 17th century. The good news is Latin can be translated into modern languages. Psychology research discovered words are learned fast by the human brain when they are associated with something familiar. Thus, assigning meaning to the Latin names makes them far easier to remember.

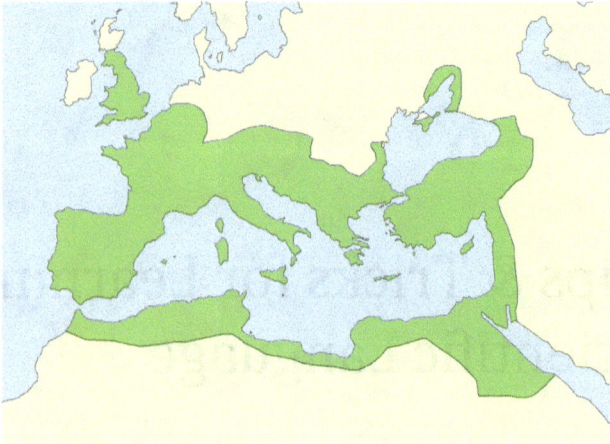

Figure: 1-1: Range of Latin language use in 60 AD shown in green. Illustration: ©Hannes Karnoefel

LANGUAGE AND SOUND

Infants and young children acquire their primary language through their brain's instinctive interpretation of auditory input. By just hearing the subset of sounds used in the language spoken near them they can sort the sounds into their proper order and map them to importance. Most brain structures dedicated to processing of auditory signals are superb at discerning pitch of the human voice and assigning implication to tones and inflection.

Infants can distinguish all of the sounds of all of the world's languages until about age six months. Between six months and a year, brain pathways devoted to language begin to form in support of the sounds most often heard. Learning to recognize and speak a language is instinctive for infants.

Adults trying to learn a second language find reading and writing a new language is not enough to develop fluency. Listening to language spoken in the correct manner over an extended period of time is needed. Auditory input is required to build new language pathways in the brain to parallel those of the native language learned in infancy.

Likewise, just reading scientific terminology does little to establish it in memory. Few people speak Classical Latin anymore, so a substitute auditory strategy is needed to help the mind map the sounds of scientific names to their meaning.

Scientific Vocabulary

Today much of the world's population is at least familiar with the English language. Some argue English should be the primary language used to teach science. And, English in its various forms is, for the most part, derived from Latin. Latin and Greek scientific words present a greater challenge for those whose native language is not derived from Latin.

Translation of compound scientific words is not always direct. The simple descriptive nature is often hidden because of the patched together arrangement of many ideas. The solution is to break the long words into parts and to assign meaning to each part. Then the parts must be rearranged into a sensible order, and word order is not always the same from language to language. For example, in Latin adjectives follow nouns unlike English where adjectives precede nouns.

Because people become so uncomfortable with the sound of scientific words, they also fail to speak and write them with precision. Scientific terminology is often composed of made-up words, which seem almost like brief descriptive pseudo-sentences. If the compound words are not spoken with precision, the various parts may become mixed in a haphazard sequence producing nonsense descriptions. To keep the parts of compound scientific names in proper order, speaking and listening must be included in the learning process.

STRATEGIES AND TACTICS

Recent studies at colleges experimented with approaches to help students learn scientific and medical terminology. Design of the education experiments relied upon conclusions of investigators who study the brain's process for learning language. Educators found reading a new and difficult word *out loud* three to five times each day for several days improved students' ability to remember the word, to spell it and to better absorb printed material using the word. Adding auditory input to reading of scientific words was more effective in creating word memory than reading alone.

The remaining sections of this chapter discuss some basic terminology needed to describe how the brain works. This vocabulary will be used often in the rest of the book. Important words will be presented in italics and the meaning of the original Latin or Greek word will be underlined.

There are online tools available for learning how to pronounce anatomic names. The tools provide an acceptable pronunciation in many native languages. An example of these tools can be found by opening a computer or tablet device to the internet at www.translate.google.com.

At Google translate, start by picking English above the box on the left and type *'neuron'* into the box. Next, to hear the word neuron in a second language, pick the second language above the box on the right. Neuron will be translated into the selected language. Below the box on the left when English is the chosen language there will be a definition of what the word means.

Below each box is a small microphone icon. Click on each icon to listen to *neuron* pronounced in the selected language. The word neuron, even though spelled the same in several languages, may be pronounced in various ways because the alphabet is pronounced in a variety of ways from language to language.

Practice pronouncing the word neuron after the computer speaks it in each language. Repeat this process three to five times for both forms of the word. The repetition will map the sequence of the sounds to memory. Keep Google translate open, and as new scientific words appear continue to practice listening to them and saying them out loud.

NAMING BRAIN ELEMENTS

Naming the cells of the brain offers a good place to begin learning how the anatomic labeling system works. For studying the brain, the scientific names *neuron, nerve cell* and *nerve* are essential. Nerve is often used as if it means the same as neuron or nerve cell. But, that is not correct. Both neuron and nerve cell refer to an individual electrical cell of the brain or spinal cord.

In contrast, a nerve is a cable-like bundle. The bundle includes just the part of a neuron called an axon. The word *axon* comes from the Greek word for axis, <u>a straight line</u>. Many neurons contribute their elongated axons to a nerve. Each axon in a nerve is the lengthy extension of a single neuron (*Figure 1-2* and *Figure 1-3*).

Nerves are enclosed by a tough sheath of tissue. The word *neuro,* from the Greek language, means <u>sinew or string</u>. Nerves in fact look like white string when seen in living tissue. The individual cells of the nervous system, neurons, were not observed by scholars until long after nerves were described (*Figure 1-3*). Some, but not all neurons, are long and stringy like nerves. Neurons assume many different shapes.

Practice reading and saying neuron, nerve and axon using www.translate.google.com.

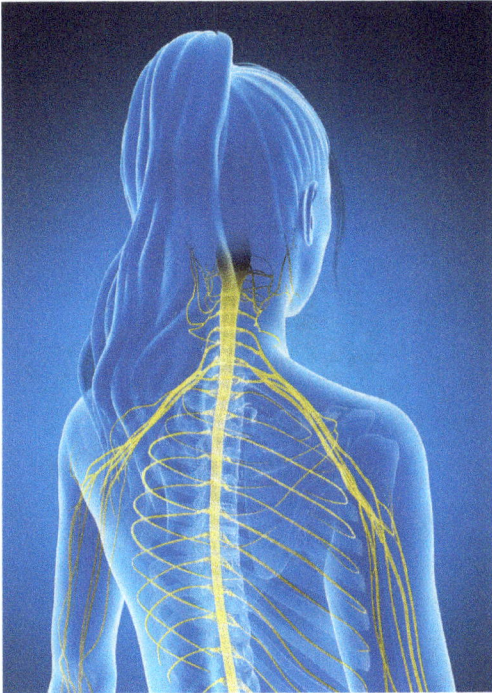

Figure 1-2: Nerves leaving the spinal cord (yellow) to head, arms and rib regions. Illustration: ©Sebastian Kaulitzki

Some neurons measure as long as three to four feet. Long neurons possess several distinct segments. One segment is the axon. Another neuron segment is the *dendrite*. A dendrite is a series of membrane projections that radiate from the body of a neuron. Dendrites divide like branches on a tree (*Figure 1-3*). The name dendrite originated in the Greek language from a word meaning tree. Practice saying and hearing dendrite and think of a neuron as having a tree like structure at one end. The word dendrite will appear often as the story of the brain unfolds.

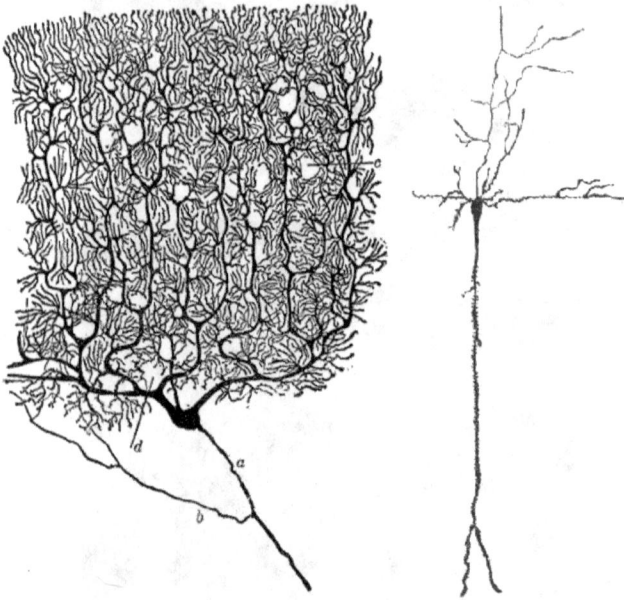

Figure 1-3: Brain neurons with different shapes. Drawing: Santiago Ramón y Cajal about 1900, this work is in the public domain

Dendrites display small membrane protrusions called *dendritic spines* (*Figure 1-4*). Here dendrite is changed to the descriptive form, *dendritic*. *Spine* is a derivative of the Latin word *spina* meaning <u>thorn-like structures on a stem</u>. Each dendrite may display several thousand dendritic spines. Dendritic spines change their shape over time in response to their local environment. Their properties draw considerable attention in modern neuroscience studies.

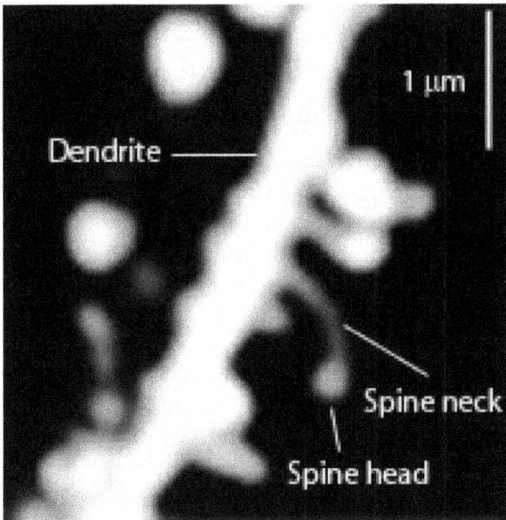

Figure 1-4: Close up of a dendrite of a brain neuron displaying dendritic spines. Photomicrograph: ©CopperKettle

Each neuron includes a *nucleus,* an area within its body to house its genetic information. As sometimes happens in anatomy, the word nucleus has two different meanings in the *central nervous system,* which is the brain and spinal cord. When describing the location of the genetic information in a neuron, nucleus means the same compartment found in other cells for housing genetic information.

But, in the brain and spinal cord *nucleus* also means a collection of neuron cell bodies. Brain areas marked by a group of neuron cell bodies fine-tune particular functional operations like fingers typing on a keyboard. Nucleus originates from the Latin word for <u>kernel</u> or <u>nut</u>, which is a suitable description of the appearance of the clusters of neuron cell bodies in the brain and of the subdivision of all cells where genetic information is stored.

The process of *neuroplasticity* is a rather new concept in science of the brain that dates back only to the 1980s. It refers to the brain's ability to rearrange its neuron dendrites and dendritic spines in response to sensory stimulation like sound and light. Neuroplasticity happens while saying, hearing and reading new scientific words. The regions of the human brain dedicated to learning new ideas are particularly busy modifying the way neurons connect.

Before 1980, scientists believed all neuron connections in the brains of mammals and birds remained permanently in place after puberty. The earliest accounts of neuroplasticity described seasonal changes in brain neuron connections in song birds. It was not until after 2000 that neuroplasticity was confirmed in human brain. Contemporary studies report human brain neurons adapt to their environment throughout life.

The word *neuroplasticity* is a combination of two words, *neuro* and *plasticity*. Plasticity originates from two similar words one Greek, the other Latin describing the process to mold. Therefore, the compound word neuroplasticity means to mold or modify how neurons connect.

Axon terminals exist at the far end of the neuron's axon (*Figure 1-5*). Axon terminals possess special characteristics allowing them to communicate with another cell. Where the terminal end of a neuron contacts another cell, a structure forms named a *synapse*. Synapse derives from a Greek word describing a <u>point of contact</u>.

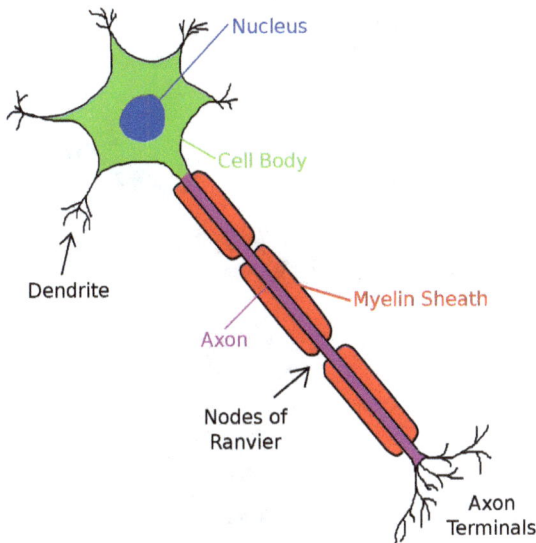

Figure 1-5: General structure of a neuron. Illustration: ©NickGorton

At a synapse an axon terminal releases a chemical substance named a *neurotransmitter* (*Figure 1-6*). Again scientists combined two words to create a new descriptive label. The word *transmitter* stems from a Latin word meaning <u>to send</u>. The combination word refers to a chemical a neuron releases as a signal.

A subdivision of neuroplasticity is *synaptic plasticity*. Synaptic plasticity is remolding of anatomic structures where axon terminals make contact, synapses. It includes changes in the type and amount of neurotransmitter released by the axon terminal. It also incorporates any modification of the receiving cell's ability to respond to neurotransmitter.

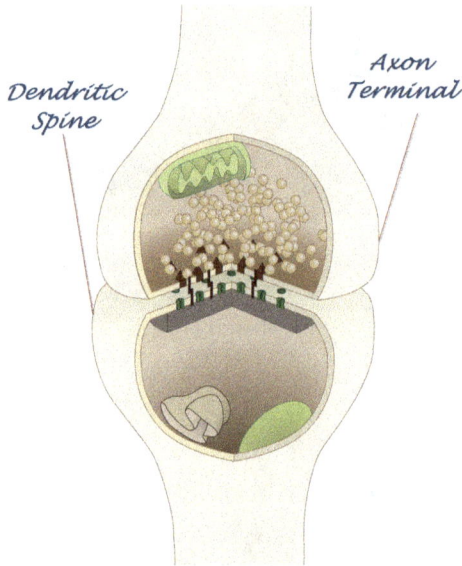

Figure 1-6: A simplified illustration of an axon terminal synapse on a dendritic spine. The beige spheres in the axon terminal represent neurotransmitter. A small space exists between the two structures through which neurotransmitter travels. Illustration: ©Curtis Neveu

Another recently recognized process for the adult brain is *neurogenesis*. In this case, two words combine to describe one process. The first part of the word, neuro, describes an electrical cell of the brain or spinal cord. The second part of the word, genesis, refers to being born. Combining the two parts into neurogenesis creates a word inferring the bringing of neurons into existence. The word *genesis* originates from the Greek word for <u>birth</u>. Genesis is a word used often in physiology. For example, *osteogenesis* is birth of new bone. Osteo is from the Greek word *osteon* meaning <u>bone</u>.

Young neurons develop from stem cells known as *neuroblasts*. Again scientists created a description from two words. The suffix blast appears again and again in physiology with various prefixes. Blast is defined as an immature embryonic stage in the development of a cell to maturity. *Blast* comes from the Greek word for <u>bud</u>. Neuroblasts are, therefore, stem cells committed to becoming neurons by proceeding through several intermediate forms like flower buds. In the adult brain, *neural stem cells* represent one of the intermediate forms of neuroblasts on their way to becoming neurons and glia.

Neurons comprise 10% of the population of cells in the brain. The remaining 90% of cells in brain tissue are named *glia* and *microglia*. The word glia is symbolic of scientists' lack of understanding of these cells until recent years. *Glia* originates from the Greek word for <u>glue</u>. Dictionaries still mistakenly describe glia as a network of branched cells and fibers gluing together the tissue of the brain and spinal cord.

Brain glia was at first divided into two classes, *microglia* and *macroglia*, based upon the physical size of the cells. That is, *small* glial cells, *microglia* and *large* glial cells, *macroglia* (*Figure 1-7*). Later, it was learned that size was not the distinguishing characteristic. Newer studies discovered microglia is not part of the glia, because it does not originate in the embryo from neuroblasts like glia, but rather from the embryo's primitive yolk sac cells.

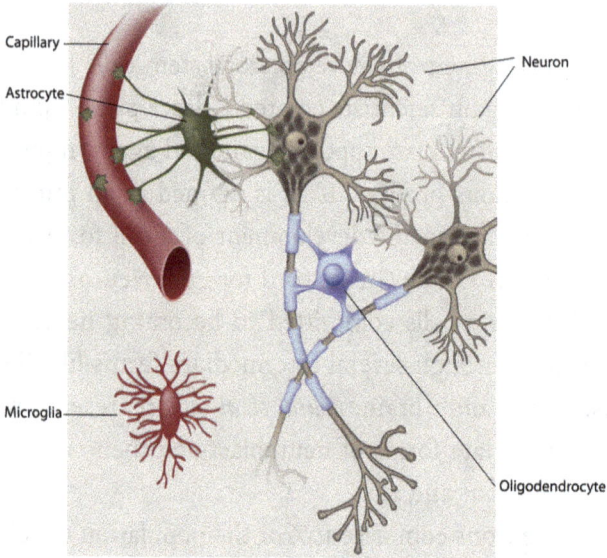

Figure 1-7: Illustration depicting four of the five types of cells in brain tissue, astrocyte, microglia, neuron and oligodendrocyte. A fifth cell type ependymal cell is not included in this illustration. Illustration: ©Alila Medical Media

Microglia migrates over a long distance in the embryo to join the neuroblasts in the developing brain. Microglia is related to *macrophages* of the body's immune system. Phage comes from the Greek word *phagos* meaning to eat. Both microglia and macrophages eat cell debris in damaged tissue. Two cell types included in glia, *astrocytes* and *oligodendrocytes* (*Figure 1-8*), develop from the same embryonic stem cell as neurons.

Brain Cell Lineage

Stem Cell

Early Progenitor Cell

Neuron   Oligodenrocyte Astrocyte

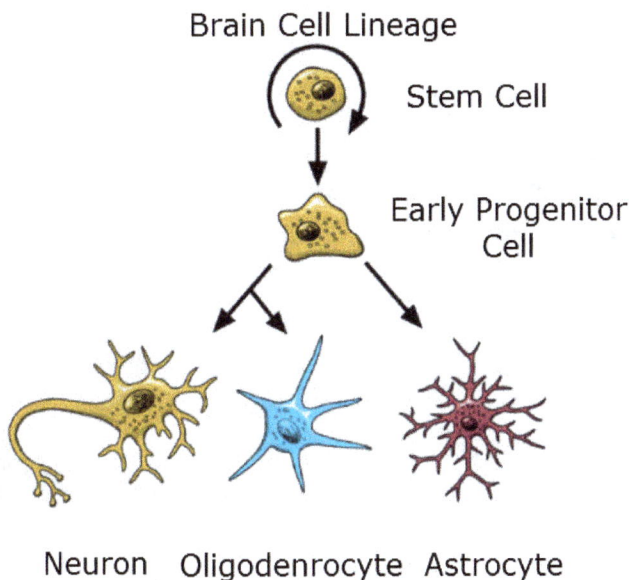

Figure 1-8: Neurons, oligodendrocytes and astrocytes all descend from the same neuroblasts, stem cells that develop into brain cells. Cell lineage but not size relationships are depicted. Illustration: This work is in the public domain courtesy of National Institutes of Health, United States

The suffix cytes is used often in physiology combined with other descriptive words. *Cytes* means cell, and it comes from a Greek word describing a hollow vessel. Cells seen with the first microscopes appeared to be hollow empty vessels.

The prefix added to cyte always describes some characteristic of the cell in discussion. *Astrocytes* appear as star shaped cells. The Greek word *astron* means a star. There are two prefixes before cyte in the case of *oligodendrocytes*. *Oligo* is a Greek word for little. *Dendro* means tree. Putting it together, oligodendrocytes translates to cells that are little and branched like trees.

The brain's ependymal cells are also derived from neuroblasts. The ependyma is a layer of cube shaped cells covering the surface of the brain's four interior chambers, the ventricles, and the central canal of the spinal cord. *Ventricle* comes from a Latin word meaning <u>belly or cavity</u>. *Ependyma* is derived from a Greek word for a <u>covering</u>.

Ependymal cells secrete *cerebrospinal* fluid which cushions the brain within the skull. The name of this fluid describes its location. *Cerebra* is a Latin word for <u>brain</u> and spina in this case refers to the hard, pointed backbone which encloses the cord of axons leaving and entering the base of the brain.

USEFUL TOOLS

For most anatomic terms, the Latin and Greek root words can be found with a little research. Often textbooks include a glossary containing some of them. Also, Wikipedia offers a helpful list of Latin and Greek root words at <u>http://en.wikipedia.org/wiki/List_of_Greek_and_Latin_roots_in_English</u>.

Another helpful reference is The American Heritage College Dictionary, published by Houghton Mifflin Company. In the dictionary, the definition of each word is followed by the Latin or Greek source word and its meaning. Discovering the sense of scientific words and hearing them spoken over and over is critical to the brain's ability to retrieve them from memory when needed.

The scientific words described here are found throughout this book and in most talks describing neuroscience. It may seem this chapter takes a long time to complete as you work with Google translate. Do not worry about it. Learning this vocabulary now will save a great deal of time later.

SUMMARY CHAPTER 1

- Human brain's ability to learn a new language is influenced by the language it learned first
- Human beings remember better new words they hear than new words they read
- Adding auditory input to reading of scientific words is an effective tool for creating word memory
- Human anatomic structures were first named by teachers who spoke Classical Latin
- Memory of words forms quicker when meaning of new words is tied to something already known
- The American Heritage College Dictionary provides Latin and Greek root words with its definition of English words
  - The Google translate website is a useful tool for practicing the language of brain science

[2]

How the Human Brain Is Organized

BRAIN STRUCTURE IS DESCRIBED in three ways. First, visual observation of the whole brain establishes the overall layout of larger structures. Second, microscopic visualization of fixed, sliced and stained brain tissue displays its cell structure. Third, videos of living brain obtained with computerized microscopes demonstrate mobility of resident neurons.

Neuron signaling practices of the human brain are more complex than those of other species. Yet, the gross organization of brain tissue is similar among mammalian species. And, a great deal of what is known about the human brain's operational systems comes from observations of rats, mice and non-human primates.

THE VISIBLE BRAIN

The human brain is a soft fragile organ protected from injury by the hard bony case of the head. Because of its soft character, the brain was considered an irrelevant organ until the late 1800s. Today scientists recognize the brain as the physical location of consciousness.

The expression *gross anatomy* refers to the external features of a dissected tissue or organ. It includes everything a person sees when viewing a body part without the help of a microscope. It may also include the texture of the tissue. For example, does the tissue feel firm or spongy?

The cow brain (*Figure 2-1*), like the human brain has a cerebellum and a right and left hemisphere. The hemispheres connect to each other by a bridge of neuron axons named the corpus callosum. Corpus callosum comes from two Latin words, *corpus* referring to a <u>body</u> of tissue and *callosum* indicating its hard texture, much like the consistency of a <u>callus</u>.

The corpus callosum appears as a broad white band of tissue composed of axons of the neurons residing in the brain hemispheres. The axons of the corpus callosum connect corresponding parts of the hemispheres, and permit the right hemisphere and the left hemisphere to coordinate their activity. The pons is a structure that attaches the cerebellum to the rest of the brain.

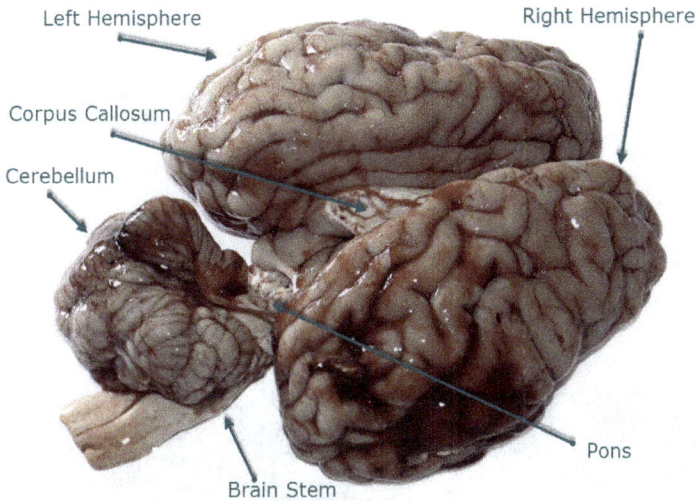

Figure 2-1: Gross anatomy of a dissected cow brain. This figure shows the placement of large brain formations. Labeled areas correspond to similar structures found in human brains. Photo: ©decade3d

In contrast to the soft consistency of dissected cow brain shown in *Figure 2-1*, a brain preserved with chemicals feels like rubber. *Figure 2-2* is a picture of a human brain treated with formalin to preserve it from decay. The increased mechanical strength of preserved brain allows the tissue to be sliced into thin sequential tissue sections. Tissue sections may be stained with various dyes to observe their cellular organization with a microscope.

Notice the deep folds in the surface of the human hemispheres (*Figure 2-2*). Increased depth of the surface folds permits greater expansion of the volume of the hemispheres without requiring the human skull to enlarge.

Similar folds of the cow brain (*Figure 2-1*) are shallow in comparison.

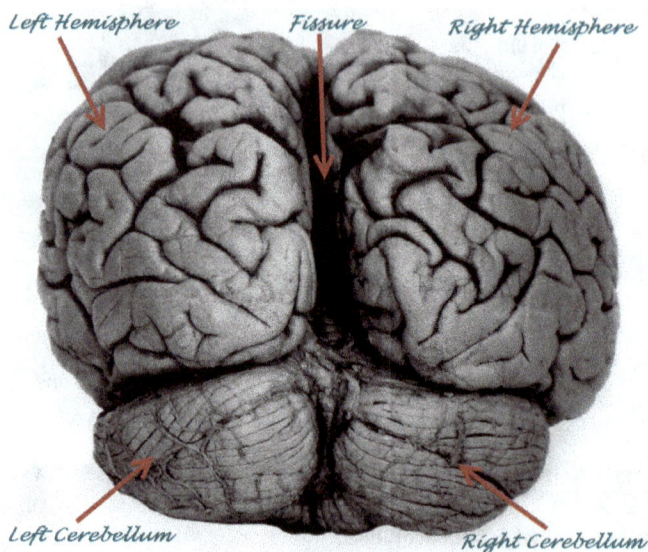

Figure 2-2: A preserved human brain photographed from the back. Photo: ©Pete Spiro

BRAIN SUBDIVISIONS

Neuroscientists of the 1800s described the human brain at birth as an organ with five subdivisions. The names assigned to the five subdivisions are based upon how the brain forms in the human embryo. Studying brain tissue in this manner proved perceptive. Neuron connections between the five subdivisions of the brain offer an anatomic framework for understanding how a brain operates. In the human embryo, the earliest neural tissue appears late in the fourth week of gestation as a hollow, fluid-filled tube with closed ends and four subdivisions (*Figure 2-3*).

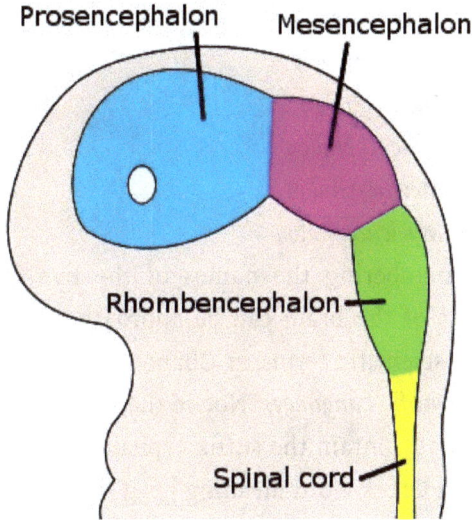

Figure 2-3: This diagram shows the position of the neural tube at about four weeks gestation in a human embryo. Illustration: ©Kurzon

The white dot shown in the blue *prosencephalon* of *Figure 2-3* develops into the optic nerve, retina and iris of the eyes. Eyes occupy a unique setting being the only parts of the brain without a bony cover.

For all mammals as the embryo matures, the pros-encephalon divides into two areas named the *telencephalon* and *diencephalon*. The embryonic subdivision identified as *mesencephalon* continues to maintain its original name even after birth. The embryonic subdivision in *Figure 2-3* labeled *rhombencephalon*, also matures into two brain subdivisions named the *metencephalon* and the *myelencephalon*. The spinal cord matures to become the adult spinal cord.

Thus, the names of the five subdivisions of the human brain at birth are:

- *Telencephalon*
- *Diencephalon*
- *Mesencephalon*
- *Metencephalon*
- *Myelencephalon*

Remembering the names of the five anatomical subdivisions of the brain can be approached in the same way as the scientific terms in Chapter 1, *"Tips & Tricks for Learning Scientific Language."* Notice the names of the brain subdivision all contain the suffix cephalon. *Cephalon* originates from a Greek word meaning <u>head</u>. Therefore, an anatomic name including cephalon indicates a part of a person's head.

Knowing the names of the brain subdivisions is useful, because neuroscientists refer to them often. For example, knowledge of the position of the brain's subdivisions would be needed to describe the path to the spinal cord of the brain's neurons devoted to initiating muscle movement.

TELENCEPHALON

The prefix of each subdivision name is descriptive of its gross anatomy. In the human brain the *telencephalon* is the right and left hemispheres, seen when looking at a whole brain as in *Figure 2-1* and *Figure 2-2*. In Greek *telos* meant far end. During embryologic development this part of the brain matures at the far end of the neural tube.

Across species the telencephalon is the brain subdivision most recently evolved. The telencephalon is required for rational thought, making decisions and implementing choices. The outermost layer of the telencephalon is named the *cerebral cortex*. *Cortex* derives from a Latin work meaning bark, as in tree bark. The cerebral cortex consists of large neurons in layers (*Figure 2-4*).

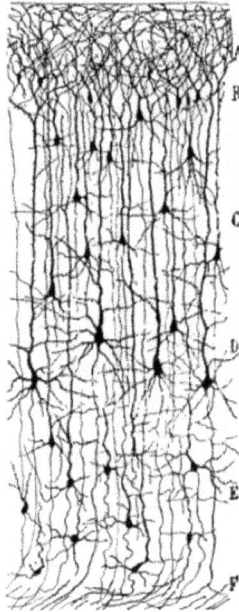

Figure 2-4: The position of cerebral cortex neurons drawn by Santiago Ramón y Cajal. Neuron cell bodies are arranged in horizontal layers labeled A-F. Axons and dendrites form a network for optimum interaction between neurons. Drawing: This work is in the public domain.

The neurons of the frontal cerebral cortex, the brain region under a person's forehead, serve as decision makers. Axons of cortical neurons are sometimes long, and many of them connect with multiple areas of the brain and spinal cord. For example, axons of cerebral cortical neurons devoted to control of body movement may extend several feet before connecting to spinal cord neurons, which in turn send their axons to muscles to cause contraction.

METENCEPHALON

Most of the *metencephalon* can also be seen by looking at the outside of a whole brain. It includes the *cerebellum* (*Figure 2-2*) and a span of tissue named the *pons* (*Figure 2-1*). *Mete* of metencephalon derives from a Latin word meaning to set a <u>boundary or limit</u>. *Cerebellum* is a diminutive form of cerebrum, a <u>small brain</u>. The original meaning of the word *pons* is <u>bridge</u>. The pons forms a physical link between the cerebellum and the telencephalon.

The surface of the brain under the forehead, the frontal cerebral cortex, is where decisions to move the body are made. The motor neurons of the motor cortex at the top of the head respond to each decision. But it is the pons and cerebellum that define the boundaries of the resulting movement by setting limits on the signals of the motor neurons before they leave the brain.

Absent the cerebellum's limiting effect on the quality of motor neuron signals, body movements lose precision and smoothness. The cerebellum is responsible for coordination of fine muscle movements and learned automatic skills including among others singing, riding a bicycle and driving a car.

Myelencephalon

The *myelencephalon* is often referred to as the *brain stem*. The brain stem is the part of the brain continuous with the spinal cord, but is within the skull. The brain stem is also called the *medulla oblongata* (*Figure 2-5*).

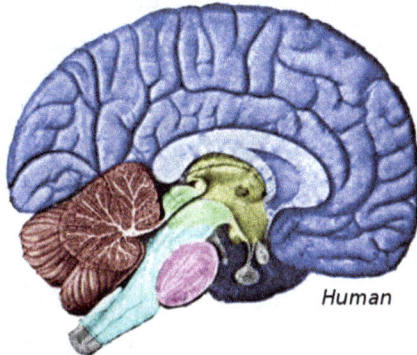

■ Telencephalon	■ Cerebellum
■ Diencephalon	■ Pons
■ Mesencephalon	■ Medulla

Human

Figure 2-5: A diagram of the medial side of the left hemisphere of a human brain displaying the location of major structures of the human brain. Illustration is in the public domain in the United States.

It serves as a communication cable composed of neuron axons coming to and leaving the brain. Medulla oblongata means a <u>long white rope</u>. This description is consistent with its appearance and with its purpose as a connecting cable. The *myel* part of the name myelencephalon refers to the fatty material wrapped around neuron axons named *myelin*.

DIENCEPHALON

The diencephalon of the human brain cannot be seen without dissection. It lies immediately below the right and left hemispheres and the corpus callosum (*Figure 2-5*). The diencephalon includes two parts. In this case the prefix *di* describes a division of the brain containing two clusters of neuron cell bodies, the *thalamus* and *hypothalamus*.

The original meaning of *thalamus* is <u>anteroom</u> or <u>entrance</u>. This is an appropriate descriptive name for this division, because almost all information coming into the brain must be processed by the thalamus before reaching other regions. The thalamus plays a central role in managing information arriving from the eyes, ears and other sensory organs.

The *hypothalamus* sits below the thalamus. The prefix *hypo* is Greek and means <u>under</u> or <u>beneath</u>. The hypothalamus controls body temperature, hunger, thirst and release of hormones from the body's master endocrine gland, the pituitary. The pituitary sits in a pocket in the bone of the skull below the hypothalamus and is connected to the hypothalamus by the pituitary stalk, a small tube of neural tissue. Pituitary is another name based upon a mistaken scientific belief. It comes from the Latin for mucus, because it was thought to be the source of mucus in the nose and sinuses.

MESENCEPHALON

The *mesencephalon* or midbrain is located deep in the center of the human brain (*Figure 2-5*). Mes is a variation of the Greek word *mesos* or <u>middle</u>. This is an old division in terms of the brain's evolution. All animals possess this brain division that co-ordinates complex reflex reactions. The mesencephalon works with the brain stem to initiate and perform the vital unconscious processes of the body like regulation of breathing. The mesencephalon is detached from intellectual reasoning.

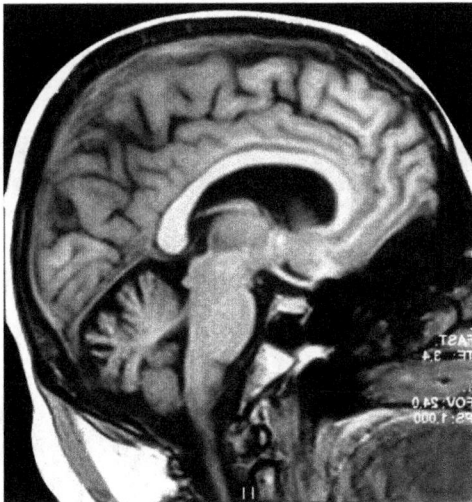

Figure 2-6: Magnetic Resonance Image (MRI) of a living human brain displaying the medial side of the left hemisphere. MRI Image: ©Cessna152

Try to identify the structures labeled in *Figure 2-5* in the image of a living human brain shown in *Figure 2-6*. Where are the cerebellum and the pons in *Figure 2-6*? Comparing *Figure 2-5* and *Figure 2-6* estimate the location of the midbrain.

In *Figure 2-6*, the corpus callosum is the white, curved structure in the middle of the image. The medulla oblongata is immediately above the spinal cord. The spinal cord is at the bottom of the image between vertebrae and is outside the skull. The scalp is the white outer line over the skull. The dark band beneath the scalp is bone of the skull. A hole in the bottom of the skull permits neuron axons to leave and enter the brain. The opening is named the *foramen magnum* and literally means <u>a large hole</u>.

GRAY MATTER AND WHITE MATTER

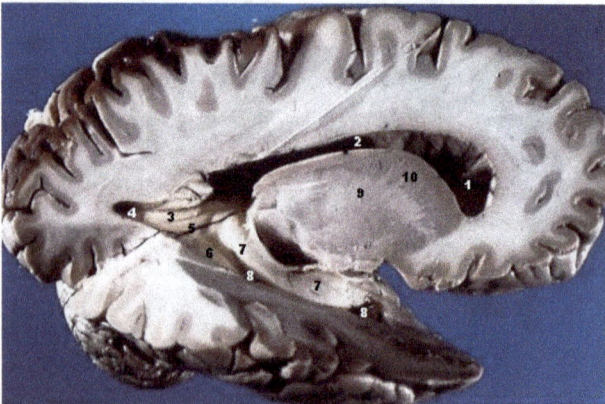

Figure 2-7: Cut surface of a fixed human brain showing gray matter and white matter. Photo: ©John A. Beal

When formalin fixed brain is sliced open, part of the interior appears white and part is a light gray color. (*Figure 2-7*). Late in the 19th century when dyes specific for neurons became available for the first time, it was discovered gray matter is clusters of neuron cell bodies. White matter is white because it contains a large number of neuron axons covered with white myelin.

INSIDE THE BRAIN

DEAD BRAIN MICROSCOPY

Microanatomy refers to anatomic features of a tissue detectable by the human eye only after magnification. The microscope was invented in the late 1500s. Cells in living tissue were first described by Robert Hooke in 1665. Yet, as late as the mid-nineteenth century some scientists still believed the brain an exception to the rule that all living tissue is made up of cells. Cells could not be seen in brain tissue because the fatty myelin interfered with dyes necessary to see the outline of the cells.

The first tissue-specific stain for neuron cell bodies was discovered in 1884 by Franz Nissl. Near the same time Carl Weigert developed a dye absorbed by the fatty myelin material of the brain and not by other brain tissue components. Comparison of these two staining methods confirmed gray brain matter contains large collections of neuron cell bodies, and white matter is white because of the large number of axons covered with myelin.

In the late 1800s and during the initial years of the 20th century, Santiago Ramón y Cajal made his first revealing drawings of neurons in brain tissue (*Figure 2-8*). The silver staining process he used was developed in 1873 by Camillo Golgi. Golgi's stain displays only a small percentage of the neurons in a tissue slice. This is fortunate because a stain that marks all of the neurons in the tissue would obscure the shape of individual neurons. Now over 150 different types of brain neurons are distinguishable based upon the shape of their dendrites alone. The many unique dendrite patterns make the neuron the most diverse cell type in the body.

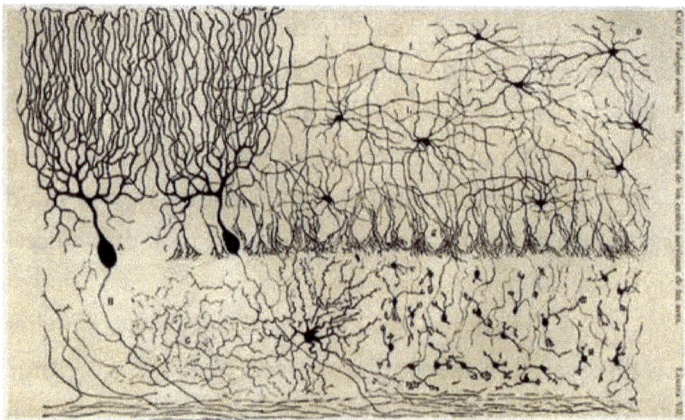

Figure 2-8: A drawing of neurons in the chick cerebellum. Notice the variety in the shape and size of the neuron cell bodies (dark round and oval structures), axons and dendrites. Drawing: Santiago Ramón y Cajal 1905. This work is in the public domain.

During the 20th century, many brain specific stains were developed for microscopic evaluation of fixed tissue slices. Modern methods permit a more detailed analysis of the cells in various brain regions. Thin pieces of tissue evaluated with contemporary techniques present a different view of the brain than observed by Santiago Ramón y Cajal.

Modern staining protocols produce images where the number and size of the brain's neurons, microglia and glia become visible. In photos taken through a microscope's magnifying lens, neurons exhibit larger cell bodies than other brain cells (*Figure 2-9*).

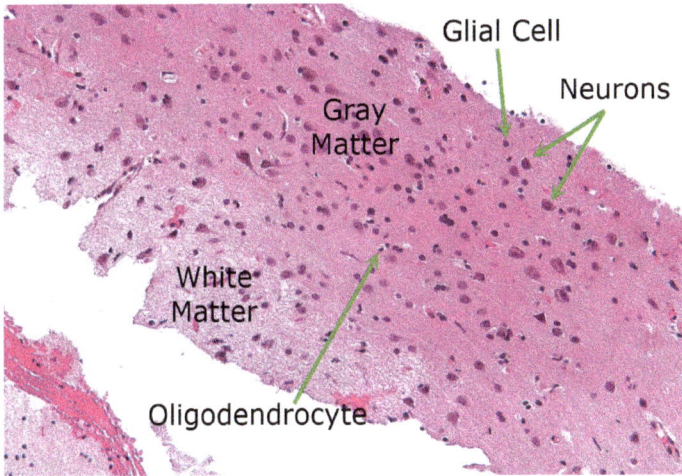

Figure 2-9: A high magnification photomicrograph of a HPS (hematoxylin phloxine saffron) stained brain biopsy. This piece of brain tissue is mostly gray matter with a small amount of white matter in the lower left quarter of the image. Photomicrograph: ©Nephron

It takes a little practice to see the difference between gray matter and white matter in stained brain sections. Areas dominated by large neuron cell bodies are gray matter. Areas where neurons are few in number are white matter. Glia is found dispersed throughout both gray matter and white matter.

Notice in *Figure 2-9* the many large neurons present in the section labeled gray matter. The cell pointed out as a glial cell is a smaller and darker staining body than the neurons. The glial cell may be an astrocyte because many astrocytes surround the neuron cell bodies of the gray matter. The smallest dark staining cells encircled by a white halo are oligodendrocytes. The halo is where fat of their myelin membrane was removed by chemicals used in the staining procedure. The part of the photomicrograph labeled white matter is primarily axons of the neurons located in the gray matter mixed with glia.

LIVE BRAIN MICROSCOPY

A method of microscopy developed since 1990 allows scientists to study a living brain. This procedure employs instruments known as optical imaging systems. Optical imaging systems provide the spatial resolution necessary to reveal individual neuron details like the shape of dendritic spines. This is an invasive procedure restricted to use in animal studies. It requires either thinning of the bone of the skull, or skull removal over the area of interest. Studies conducted in mice, rats, cats and non-human primates provide most of these data.

When using rats, a permanent window can be implanted where the skull is removed and imaging living neurons can be performed for a year or more in the same animal. Mini portable devices allow imaging of neurons to occur while rats explore their environment.

At present the finest optical brain images penetrate to a depth of about 1 millimeter of the brain surface. For this type of optical imaging of neurons to be useful in humans, the depth of light penetration must be improved, and a way to avoid an open skull must be found. Improvement of optical imaging technology is a goal of the present worldwide emphasis on brain research.

In reality the picture detected by optical imaging systems, confocal microscopes and 2 photon imaging microscopes, cannot be seen directly by human eyes. These are not microscopes in the same sense Robert Hooke's instrument is a microscope. Light information from modern imaging systems is sent to a computer and the computer forms a picture that the human eye recognizes (*Figure 2-10*).

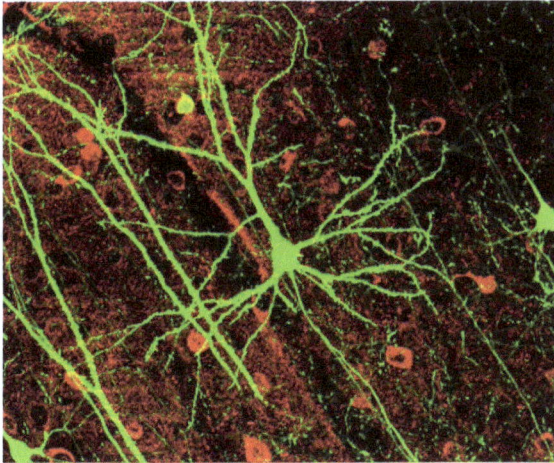

Figure 2-10: A pyramidal neuron expressing Green Fluorescent Protein (GFP) in a mouse visual cortex. Photomicrograph: ©Nrets

Light captured from brain tissue by optical imaging systems is produced by fluorescent molecules. A focused laser beam is used to increase the energy level of the fluorescent molecules in the tissue with fast pulses of infrared light. Between pulses the fluorescent molecules return to their normal energy state. In the process of returning to their baseline energy level, each fluorescent molecule emits a photon of light at a particular wavelength in the visible light range.

The amount of light produced by fluorescent molecules is so low it must be enhanced as part of the detection process. Thus, a computer is required to compile into a visual image the light emitted, the light scatter information and the position of the focused laser beam in the tissue.

For brain cells to possess fluorescent molecules, the animal must make them by using its own cell machinery for synthesizing molecules. Animals modified to do this are named *transgenic animals*. A common way to produce transgenic animals is to inject the genes required for synthesis of the fluorescent molecule into the nucleus of a fertilized egg.

Correct timing of the steps of the procedure is essential. For a foreign gene to be incorporated into an animal's genetic material, the alien gene must be added before the chromosomes of the sperm and oocyte merge (*Figure 2-11*). In most experiments, up to 40% of the mice born from such embryos will express the foreign gene and make fluorescent protein. Expression of the foreign gene can be restricted to a particular tissue in the animal by including a part that responds to molecules unique to the tissue.

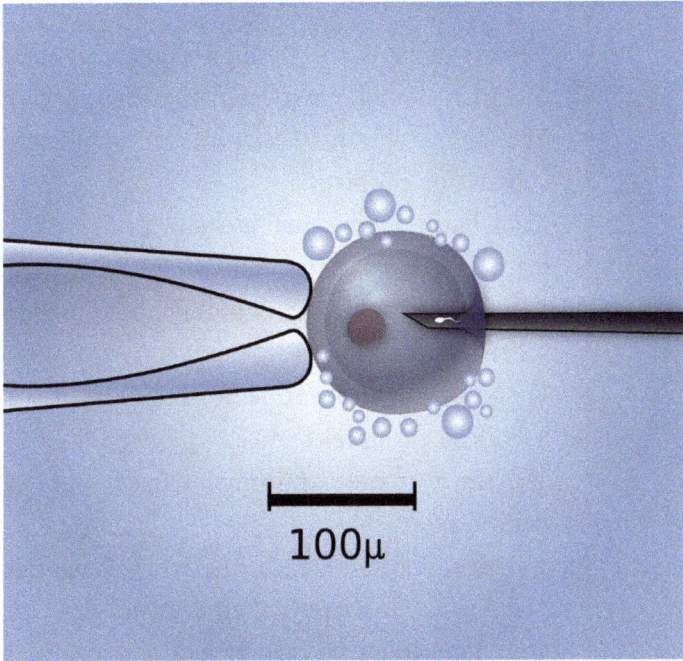

Figure 2-11: Diagram of sperm injection into an oocyte to create a fertilized egg. The micro-manipulator holds the oocyte while the micro-injector on the right places a single sperm in the oocyte. Illustration: ©KDS444

Computerized microscopic techniques not only permit observation of brain anatomy in greater detail but also display the animated nature of brain tissue. Videos made by Michael J. Schell http://youtu.be/Mhcaz6_fzZA and http://youtu.be/Cjjdky96ubc demonstrate the motion of dendritic spines in real time.

SUMMARY CHAPTER 2

- The organization of the major subdivisions of the brain is similar in most mammalian species
- The human brain develops into 5 anatomic subdivisions by birth
- Each brain subdivision includes a distinctive feature reflected in its name
- Resident neurons connect brain regions to each other with their axons
- The gray matter of formalin fixed brain tissue contains neuron cell bodies
- The white matter of formalin fixed brain is neuron axons
- Neurons distribute into anatomic categories based upon the shape of their dendrite
- Over 150 different types of neuron categories exist based on the shape of dendrites
- Optical imaging systems for living animals deliver the spatial resolution necessary to reveal individual neurons, their extensions and over time their mobility

[3]

Quality Control of Brain's Extracellular Fluids

THE HUMAN BRAIN IS ISOLATED from the rest of the body in multiple ways. Its cells manage their business like members of an independent society. The brain depends upon the rest of the body only for an adequate supply of oxygen and glucose and a small, select group of nutrients and growth factors. It connects to the outside world primarily through its own neuron-based sensory systems.

Physiologic mechanisms adapt to the brain's unique circumstances. One variation of normal physiology is revealed in the unusual characteristics of the brain's maintenance program for its fluid compartments. Three fluid compartments support brain cell activities. They are the intracellular fluid named *cytoplasm* and extracellular fluids known as *interstitial fluid* that surrounds blood vessels, neurons, glia and microglia and the *cerebrospinal fluid* that cushions the brain within the skull.

While exchange of molecules between fluid compartments is a dynamic process in all body tissues, brain tissue exhibits an unusual and elegant form of molecular exchange between its fluid compartments. This chapter focuses upon quality control of the brain's extracellular fluids. The following chapters explain how dynamics between extracellular and intracellular fluid compartments support, and are vital to, the electrical signaling systems in the brain.

Fluid Surrounding Cells

Virchow-Robin Space

Interstitial fluid surrounding brain cells and blood vessels consists of water with dissolved sugars, salts, fatty acids, amino acids, hormones, neurotransmitters and the water soluble waste products generated by cell activity. Larger water-soluble molecules including proteins are absent in normal circumstances.

Unlike elsewhere in the body, interstitial fluid around neurons and glia is separated from interstitial fluid around the brain's arteries and arterioles by *pia mater*, an ultra-thin membrane. The name pia mater comes from Medieval Latin and translates into <u>tender mother</u>. The space created by the presence of the pia mater around arterial vessels is known as the *Virchow-Robin space*.

Virchow-Robin is a composite of the names of the two investigators who demonstrated existence of this fluid space in the brain, Rudolph Virchow and Charles-Philippe Robin. No comparable pia mater sheath is present around the veins. The Virchow-Robin space ends where blood capillaries begin. There it is eliminated by fusion of the capillary endothelial cells with the membrane of astrocyte glial cells.

Virchow-Robin space isolates interstitial fluid from all proteins and other large molecules leaked from arterial blood vessels. Virchow-Robin space also helps to clear waste-containing interstitial fluid surrounding neurons and glia. Waste-containing interstitial fluid flows through the pia mater into the Virchow-Robin space and then drains into the lymphatic system of the head and neck (*Figure 3-1*) for return to the heart.

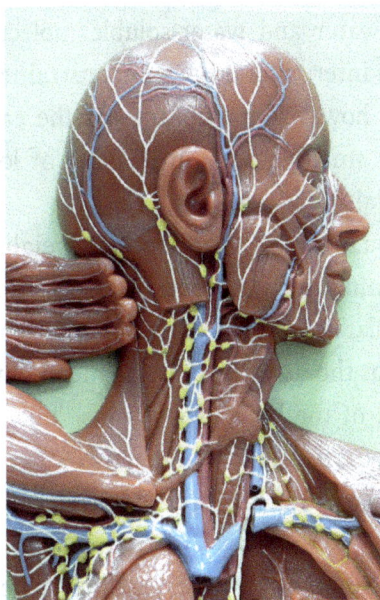

Figure 3-1: Anatomic model of the veins and lymphatic vessels of the head and neck. Photo: ©Tinydevil

Lymphocytes, white blood cells of the immune system, escaping from blood vessels become trapped in the Virchow-Robin space and are returned to the blood. In healthy brain, lymphocytes and other cells of the peripheral immune system are excluded from the interstitial fluid around neurons and glia. Only when the brain's own immune-like cells, the microglia, become overpowered by infection or trauma do immune system lymphocytes enter into interstitial fluid surrounding neurons.

BLOOD BRAIN BARRIER

Blood in vascular vessels, arteries and veins, is often included in the normal description of the body's extracellular fluid compartments. Elsewhere in the body, direct exchange of water and water-soluble molecules between blood and the interstitial fluid is unconstrained at capillaries. In brain, however, membranes of the capillaries and post capillary small veins limit passage of low molecular weight material and water.

The quantity of low molecular weight substances, hormones, amino acids, neurotransmitters and other metabolites oscillates in blood under normal circumstances. Fluctuations in the quantity of these metabolites in the interstitial fluid of the brain would cause unacceptable disruption of neuron function. About 98% of blood's small molecules do not enter the brain through its capillary system.

The limitation on release of low molecular weight material from blood increases osmotic pressure in brain capillaries. Brain capillary *osmotic pressure*, a <u>force pulling water</u> into capillaries created by the high number of molecules unable to leave the blood is great. Far too much water would be removed from the brain's interstitial fluid without a reduction in permeability of brain capillaries for water. Entry of water and molecules necessary for brain well-being uses a different path that is described below.

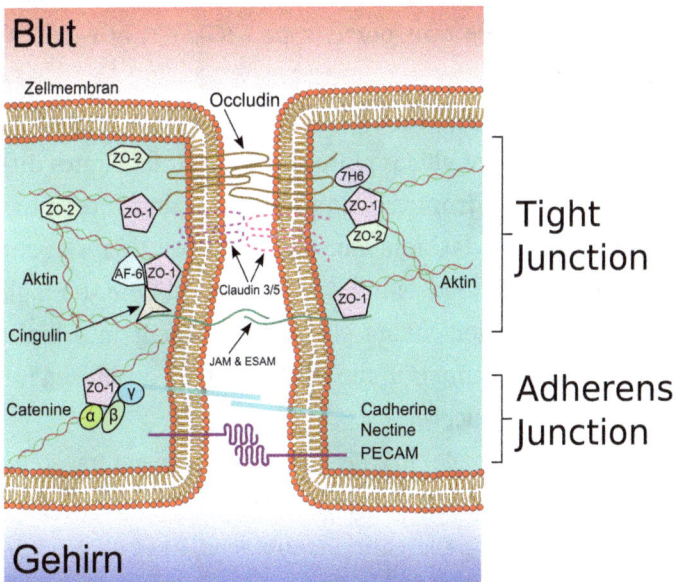

FIGURE 3-2: Diagram showing tight junctions at the blood brain barrier including some of the main proteins. Blut = blood, Gehirn = brain, Illustration: ©Armin Kübelbeck

A *blood brain barrier* is created by tight lacing together of capillary endothelial cells by protein formations with the descriptive names *adherens junctions* (*Figure 3-2*).

The junctions cause molecules in blood to be transported through the capillary endothelial cells.

The blood capillary membrane is composed of a cell type named *endothelial*. Endothelial cell membranes have proteins that transfer specific molecules from one side to the other. About 10-15% of the proteins in the brain's capillary system transport molecules. Glucose, the brain's main energy source, is transported into the brain by a protein named the glucose transporter type 1 (GLUT1) of capillary endothelial cells.

GLUT1 does not require energy or insulin to perform its transfer of glucose into the brain. It facilitates diffusion of glucose from the blood to the brain's interstitial fluid in a passive fashion. Glucose is almost always about twice as high in blood as in brain interstitial fluid. Its high blood concentration favors glucose movement into brain tissue through channels in the GLUT1 protein optimized for the structure of glucose.

Insulin made in the pancreas also enters some brain regions by using an endothelial cell transporter protein. In the brain pancreatic insulin regulates cell growth. Pancreatic insulin is reported to support survival of neurons without affecting their use of glucose. The neurons of the cerebral cortex, hypothalamus and cerebellum are particularly sensitive to insulin. Insulin supports brain uptake of the amino acids needed to synthesize the neurotransmitters norepinephrine, dopamine and serotonin.

Gas exchange is not affected by the brain's capillary barrier. Because both oxygen and carbon dioxide dissolve in the fatty matrix of cell membranes, they pass with ease through the brain's capillaries. Other small fatty molecules also pass through the capillary membrane of the blood brain barrier. However, many of them are transferred back into the blood by membrane proteins.

An additional layer of physical and operational support for the blood brain barrier comes from membrane projections of another brain cell called an astrocyte (*Figure 3-3*). The astrocyte's foot-like structures make direct contact with brain capillaries. Recent studies produced data suggesting astrocytes, in addition to providing mechanical support for the blood brain barrier, also participate in secretion of chemical factors to block immune cells from entering the brain.

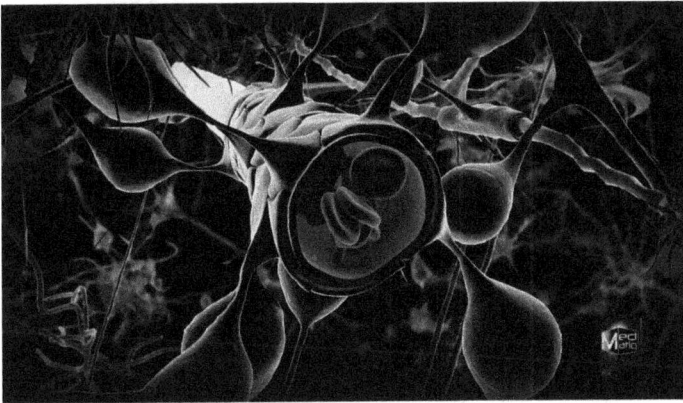

Figure 3-3: An artist's rendition of a blood brain barrier capillary with attached astrocytes. Illustration: ©Ben Brahim Mohammed

The blood brain barrier's most important purpose is prevention of entry of bacteria, virus, immune cells and large proteins like antibodies. Immune system cells enter the brain only when its blood vessels suffer damage. The presence of immune cells in tissue leads to inflammation. Typical inflammation is not well tolerated by brain tissue because it makes capillaries leaky permitting passage of water and other blood borne molecules.

THE MENINGES

The *meninges* are layers of membranes covering the entire surface of the brain and spinal cord. Their name is derived from the Greek word *meninx* meaning membrane. The meninges provide mechanical protection for brain tissue, furnish a pathway for flow of cerebrospinal fluid to cushion the brain and spinal cord and supply support for blood vessels entering and leaving the brain. Formation of cerebrospinal fluid and its passage through the brain is covered in the following section.

Three layers of meningeal membranes, dura mater, arachnoid membrane and pia mater cover the brain (*Figure 3-4*). The name *dura mater* is from Greek, and it translates to tough mother. Dura mater is a thick fibrous tissue. There are two layers of dura mater, a layer near the bone of the skull continuous with the bone's own membranous wrap and an inner layer closer to the brain. The two layers of dura mater enclose and support the large venous channels and sinuses that return blood to the heart.

Dura mater also forms a sac around the next layer of membrane, the *arachnoid membrane*. The arachnoid membrane is thin transparent fibrous tissue. It acquired its name because of the spider web appearance of its delicate fibers that connect it to the layer of pia mater underneath. The Greek word for <u>spider</u> is *arachne* and the suffix *oid* means <u>in the image of</u>.

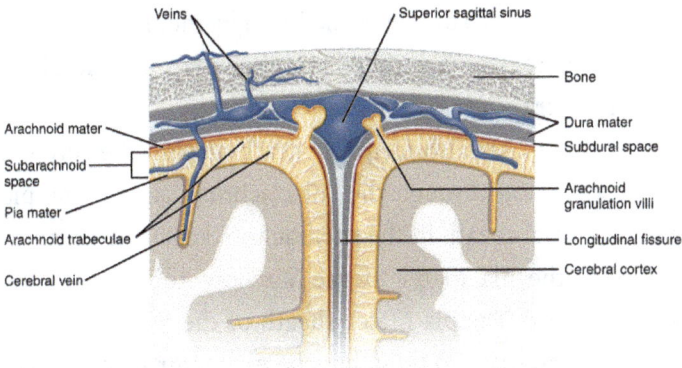

Figure 3-4: This diagram represents a section across the top of the human skull. It describes the arrangement of the meningeal membranes covering the brain. The orange area, the subarachnoid cavity fills with cerebrospinal fluid. This figure is based on plate 769 from Gray's Anatomy. Illustration: ©OpenStax College

Between the arachnoid membrane and the next layer of meninges, the pia mater, is a webbed space filled with cerebrospinal fluid, the subarachnoid cavity. The subarachnoid cavity receives cerebrospinal fluid flowing out of the fourth ventricle. The subarachnoid cavity encircles the entire brain and spinal cord providing a pad of protection. All blood vessels entering the brain, branches of the internal carotids and vertebral arteries, pass through the subarachnoid cavity. Cranial nerves exiting the bottom of the brain also passage through the subarachnoid cavity.

Pia mater is the innermost membrane and it adheres to the cerebral cortex running down into the surface fissures. At the surface fissures, pia mater is a thin fibrous tissue that is impermeable to fluid. It forms a sheer translucent envelope spanning almost the entire brain. Pia mater is anchored to the surface of the brain by membrane extensions of astrocytes like those found reinforcing the blood brain barrier (*Figure 3-3*).

Pia mater also forms a sheath around the cerebral arteries passing through the subarachnoid cavity. The pia mater arterial sheath in the subarachnoid cavity is continuous with the pia mater of the Virchow-Robin space described above. The pia mater sheath of the Virchow-Robin space is permeable to fluid, but the pia mater sheath of the subarachnoid cavity is watertight supplying a barrier between cerebrospinal fluid exiting into the venous sinus and incoming blood vessels.

CEREBROSPINAL FLUID

FLUID PRODUCTION

Cerebrospinal fluid is an extracellular fluid exclusive to the brain. It travels through the brain's inner chambers and around the outside of the brain and spinal cord. Flow of cerebrospinal fluid compensates for the limited permeability of the brain's capillary system. It delivers water and nutrients to, and removes waste products from, the interstitial fluid surrounding neurons and glia.

A small number of immune system lymphocytes populate human cerebrospinal fluid. The class of lymphocyte present suggests they may help detect the initial stages of pathogen infection. Thus, it is probable cerebrospinal fluid's resident lymphocytes perform surveillance services rather than participate in an inflammatory immune response.

Cerebrospinal fluid is continuously secreted into the brain's four inner chambers, the ventricles (*Figure 3-5*), by ependymal cells. Ependymal cells are small cuboidal ciliated cells lining the surface of the ventricles.

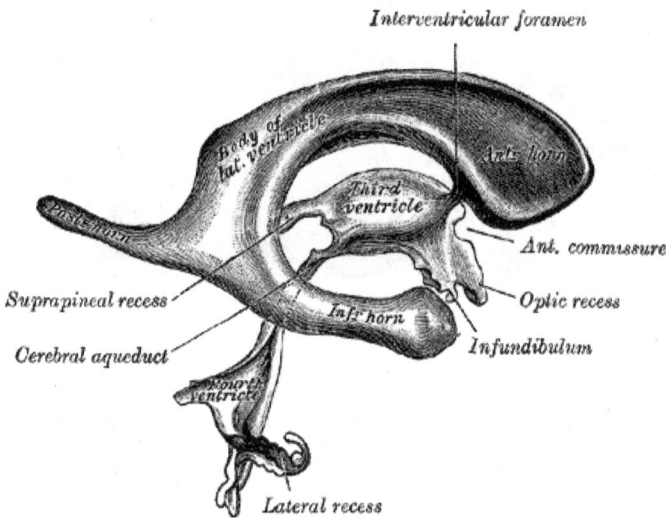

Figure 3-5: Diagram of the space occupied by the ventricles of the human brain. Illustration is in the public domain in the United States.

Continuous with the ependymal cell layer of the ventricle walls is a group of ependymal-like cells. The ependymal-like cells cover a special tissue located in each of the four ventricles named a *choroid plexus* (*Figure 3-6*). The choroid plexuses permit passage of water, selected small molecules, growth factors and nutrients from blood into the cerebrospinal fluid for delivery to the interstitial fluid.

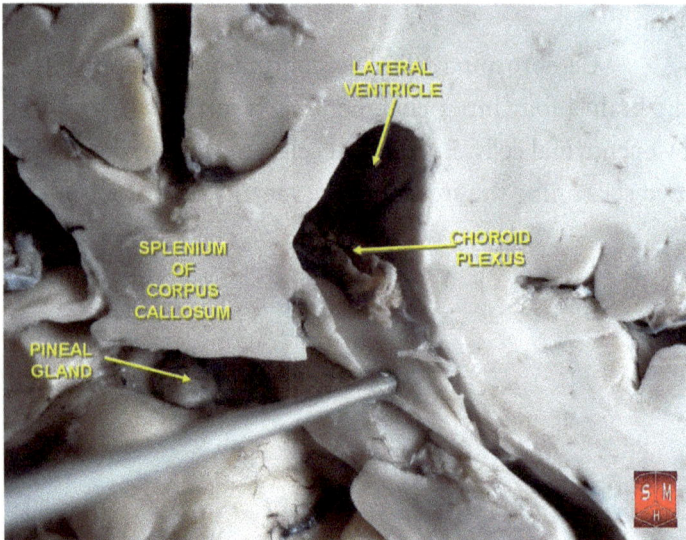

Figure 3-6: This lateral ventricle of a dissected human brain shows a choroid plexus. Photo: ©Anatomist90

The interior of a choroid plexus is a convoluted vascular network of loose connective tissue and large capillaries. These capillaries possess a structure similar to small perforated veins. Most blood-borne small molecules pass through and around the endothelial cells of these capillaries bringing water with them into the fluid within the choroid plexus. The ependymal-like cells covering the choroid plexus display extensive folds on their side facing the blood vessels. Creases in the membrane expand the cell's ability to absorb fluid released from the sieve-like capillaries.

Fluid within the choroid plexus must be processed through the ependymal-like cells to become cerebrospinal fluid. Tight junctions tie together ependymal-like cells of the choroid plexus preventing fluids from going around them. Like other cells lining hollow spaces, the characteristics of their cell membrane on the side facing the blood capillaries, their *basal membrane,* differ from characteristics of their membrane facing the open space of the ventricle. The part of their membrane facing the open chamber is called their *apical membrane.*

Proteins in the apical membrane transfer ions from the ependymal-like cell's cytoplasm into the cerebrospinal fluid. The ions create an osmotic pressure in the cerebrospinal fluid. When sufficient osmotic pressure is generated by the ion transfer process, water is pulled out of the cells into the cerebrospinal fluid increasing its volume. Nutrients and growth factors picked up by the basal membrane passage through the cells before following water into the cerebrospinal fluid. Ependymal cells also manufacture and secrete a wide variety of biologic substances to support brain health.

In addition to secreting cerebrospinal fluid, the choroid plexus is active in the clearance of drugs and pollutants from the brain. In a reverse process, under appropriate circumstance, the ependymal-like cells reabsorb cerebrospinal fluid. Reabsorbed cerebrospinal fluid moves back into the choroid plexus capillaries.

The ependymal cells lining the wall of the ventricles also absorbs cerebrospinal fluid, but for a different purpose than the choroid plexus. Along the surface of the ventricles absorption of cerebrospinal fluid is a mechanism for exchanging nutrients in cerebrospinal fluid for cellular waste products in interstitial fluid.

In humans about 500 milliliters of cerebrospinal fluid is produced each day. The volume of the entire system of channels for cerebrospinal fluid is only 150–270 milliliters. Therefore, cerebrospinal fluid is replaced about 2–4 times per day. Pressure generated by the continuous production of cerebrospinal fluid causes it to flow through the ventricles, the spinal cord and membranes surrounding the brain and spinal cord.

Path through the Central Nervous System

To illustrate the path taken by cerebrospinal fluid through, around and out of the brain, the anatomy of the system must be reviewed. The central portion of the brain is occupied by lateral spaces in the hemispheres named the lateral ventricles and two other open areas named the third ventricle and the fourth ventricle (*Figure 3-5*). An open area, the central canal, also runs the length of the spinal cord.

Cerebrospinal fluid formed in the two lateral ventricles passes through the *interventricular foramen*, a small open area connecting the ventricles, into the third ventricle. Refer back to *Figure 3-5* to see the location of the open spaces included in this path.

The third ventricle sits between the thalamus of the right hemisphere and the thalamus of the left hemisphere. There cerebrospinal fluid of the lateral ventricles mixes with cerebrospinal fluid formed in the third ventricle. From the third ventricle the combined cerebrospinal fluid of three ventricles moves through the *cerebral aqueduct* into the fourth ventricle.

The fourth ventricle lies between the right and left pons and cerebellum. Merged cerebrospinal fluid of the four ventricles exits the fourth ventricle and enters into the central canal of the spinal cord and into subarachnoid cavity of the meninges.

Cerebrospinal fluid flow from the ventricles reaching the top of the head within the subarachnoid cavity passes through structures in the arachnoid membrane named *arachnoid granulations*. It flows through the arachnoid granulations and mixes with blood in the venous *superior sagittal sinus* (Figure 3-4).

The arachnoid granulations act as one way valves for flow of cerebrospinal fluid out of the brain. Pressure in the cerebrospinal fluid is usually higher than pressure in the venous sinus. But, even when the pressure difference is reversed no back flow of venous blood occurs through arachnoid granulation valves into the subarachnoid cavity because of the structure of the valves.

CEREBRAL BLOOD SUPPLY

ARTERIAL

Although the blood brain barrier limits flow of many materials from blood into the brain, a steady cerebral circulation to deliver glucose, oxygen and a small set of other molecules is critical for brain tissue survival. Neurons require a constant source of glucose for energy production. Structures inside neurons named *mitochondria* synthesize mobile high-energy molecules of adenosine triphosphate (ATP) from glucose. They depend upon a steady supply of oxygen to support their production of ATP. Interruption of oxygen flow to brain tissue, and therefore synthesis of ATP, for as little as four minutes can cause permanent neuron damage.

Two major sets of arteries on the right and left side of the body provide blood that is rich in oxygen and glucose to the brain. They are the internal branch of the *carotid arteries* of the neck and the *vertebral arteries* (*Figure 3-7*).

The external branch of carotid arteries supplies blood to the face. The internal carotid arteries perfuse the front and middle portions of the brain and the vertebral arteries perfuse its back portion. The brain's arteries spread over its surface within the meninges in the subarachnoid space before they penetrate deep into brain tissue.

The internal carotid arteries run deep in neck tissue to the right and left of the trachea. They enter the skull through the carotid canals of the skull's temporal bone.

There they branch into the anterior cerebral arteries and the middle cerebral arteries.

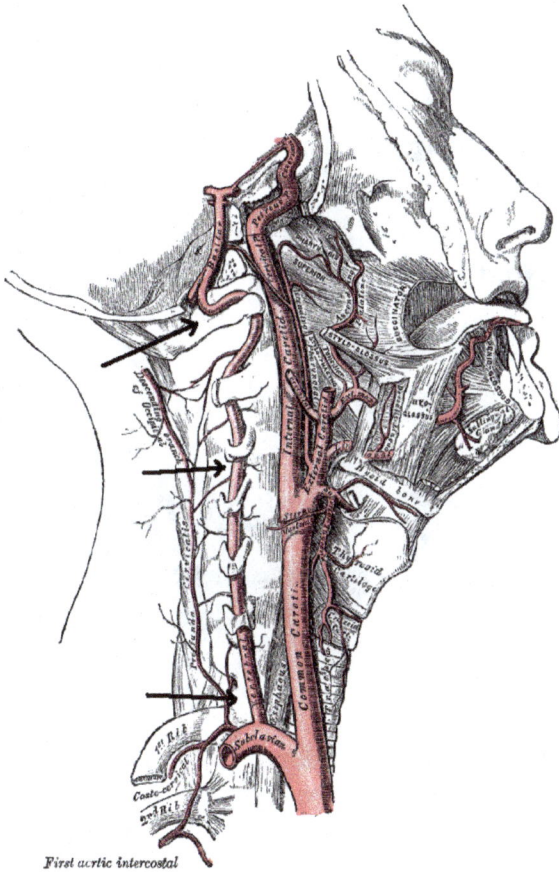

Figure 3-7: Arteries to the brain shown on the right side of the head. This illustration is a reproduction of a lithograph plate from Gray's Anatomy, published in 1918. This work is in the public domain in the United States.

The vertebral arteries are smaller than the internal carotid arteries. The pair of vertebral arteries branch from the large arteries supplying the shoulders, lateral chest and arms. They run through the lateral holes in the transverse process of the cervical vertebrae C6 to C1 (*Figure 3-7*). They then travel across the cervical vertebrae C1 and enter the brain through the foramen magnum in the base of the skull. Within the skull they fuse together to make the basilar artery that supplies blood to the midbrain. The basilar artery branches further to traverse the posterior part of the brain.

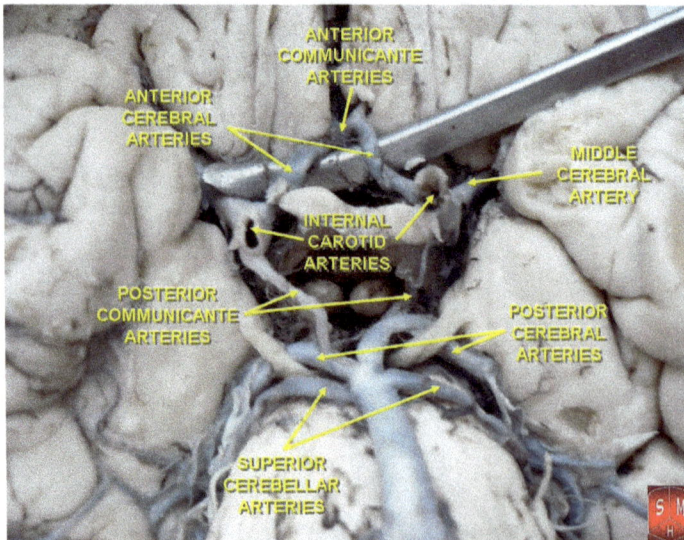

Figure 3-8: Anterior and posterior cerebral circulations interconnected by posterior communicating arteries form the Circle Of Willis beneath the brain. Photo: © Anatomist90

The anterior carotid circulation and posterior vertebral circulation connect to each other by the posterior communicating arteries at the *Circle of Willis*. The Circle of Willis lies at the base of the brain (*Figure 3-8*). This arrangement of connecting arteries acts as a safety net. If one part of the cerebral circulation becomes injured or blocked, blood flow from the other vessels can be shunted through the Circle of Willis to preserve perfusion of most of the brain.

VENOUS

The venous return from the deep brain capillary beds is composed of traditional veins. This venous system merges to form the vein of Galen behind the midbrain. The vein of Galen joins the superficial venous system composed of the venous sinuses (*Figure 3-4*) in the dura mater of the meninges.

Superficial veins draining the anterior brain also empty into the network of sinuses in the dura mater of the meninges. The right and left dura sinuses come together in the posterior brain and leave the skull as the internal jugular veins. The internal jugular veins run parallel with the carotid arteries back to the vena cava and heart.

The venous return from the posterior surface of the brain and from the cervical spinal cord travels to the heart by way of the vertebral veins and the large veins of the chest. The vertebral veins descend from the head alongside the vertebral arteries within the holes of the transverse process of the cervical vertebrae.

SUMMARY CHAPTER 3

- Interstitial fluid around neurons and glia is separated from interstitial fluid around the brain's arteries and arterioles by pia mater, a thin membrane creating the Virchow-Robin space

- Virchow-Robin space assists in blocking access to the brain of lymphocytes and proteins of the body's immune system

- The blood brain barrier prevents direct exchange of water and most small molecules between brain capillaries and interstitial fluid

- The blood brain barrier is created by tight connections between capillary endothelial cells and is supported by astrocyte membranous extensions

- The meninges are layers of membranes covering the entire surface of the brain and spinal cord

- Cerebrospinal fluid is secreted by ependymal-like cells of the choroid plexuses located in each of the brain's four ventricles

- Cerebrospinal fluid delivers nutrients, growth factors and water from blood to the interstitial fluid of the brain and removes cellular waste material

- Cerebrospinal fluid flows from the lateral ventricles into the third ventricle and then into the fourth ventricle

- Cerebrospinal fluid flows from the fourth ventricle into the central canal of the spinal cord and into the subarachnoid cavity of the meninges
- In the subarachnoid cavity, cerebrospinal fluid flows to the top of the head where it passes through one-way valves into the large venous sinuses
- A continuous flow of blood to the brain for delivery of glucose and oxygen is essential
- Major arteries supplying oxygenated blood to the brain are the internal branch of the carotid arteries and the vertebral arteries located on each side of the body
- The internal carotids perfuse the anterior and middle portions of the brain and the vertebral arteries perfuse its posterior portion
- Veins draining blood from the anterior brain empty into the network of sinuses in the dura mater of the meninges
- The dura sinuses merge in the posterior brain and leave the skull as the internal jugular veins
- The venous return from the posterior surface of the brain and from the cervical spinal cord travels to the heart by way of the vertebral veins and the large veins of the chest

[4]

Neurons—How They Make Electricity

THE LARGER, VISIBLE ASPECTS of brain anatomy and physiology are described in previous chapters. Here the narrative shifts to the realm of the less visible aspects of the brain, its electrical currents.

Neurons, the principal electrical cells and the unmistakable superstars of the brain cell community, occupy the central hub of almost all efforts to clarify how the brain works. These remarkable cells form the physical substance of the mind and a person's sense of self.

Brain's form of electricity is probably the most difficult concept for people new to science to get their mind around. It is not the same as the electricity encountered in everyday life, and it is human to doubt ideas that are unexpected and without precedent.

Yet, the seemingly bizarre explanation of neuron electricity presented here is the net result of over 40 years of scientific experimentation. And, the report is not yet complete. Neuroscientists continue to fine-tune their understanding of how neurons communicate.

Because neuron communication is complex, this chapter reviews only the electrical properties of neuron axons. The next chapter considers synapses, the interface neurons use to communicate with each other and the effect of synapse activity on dendrites and events within the body of a neuron.

NEURON COMPARTMENTS

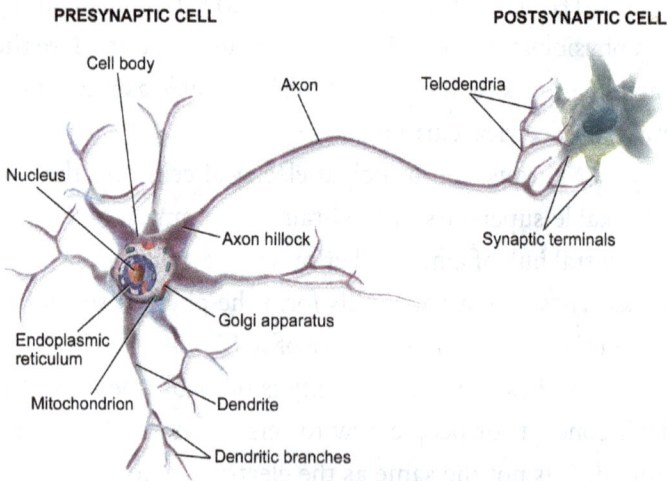

The Anatomy of a Multipolar Neuron

Figure 4-1: Artist's representation of the anatomy of a generic neuron. Illustration: ©BruceBlaus

Neurons can be dissected and analyzed as distinct anatomic and functional compartments. *Figure 4-1* is a drawing illustrating some of the compartments of a neuron including dendrites, neuron cell body, axon hillock, axon, axon telodendria and synaptic terminal. Sometimes scientific articles give the impression each part of a neuron is an independent entity. Yet neuron compartments do not act solo. Rather, each compartment serves a special aspect of one complete operation.

Common practice in teaching anatomy and physiology is to use cartoon illustrations, similar to the one of a neuron shown in *Figure 4-1*. Illustrations, while useful tools, merely approximate the genuine entity. Drawings of neurons oversimplify shape of the dendrites. They do not suggest the actual diversity of axon characteristics.

Real axons may be several feet long or they may be short and difficult to distinguish from dendrites. Real axons often split and send their divisions called *axon collaterals* to connect with multiple neurons. Axon collaterals often reach back to connect with their own dendrites and neuron body.

Yet, cartoon illustrations can orient discussions. For simplicity's sake, this chapter's examination of a single neuron's physiology includes reference to several cartoon illustrations including *Figure 4-1*. Photos of real neurons seen through a microscope are included when available.

Compare *Figure 4-1* with *Figure 4-2*. *Figure 4-2* is a picture taken through the lens of a microscope of brain tissue stained to show part of the structure of real neurons.

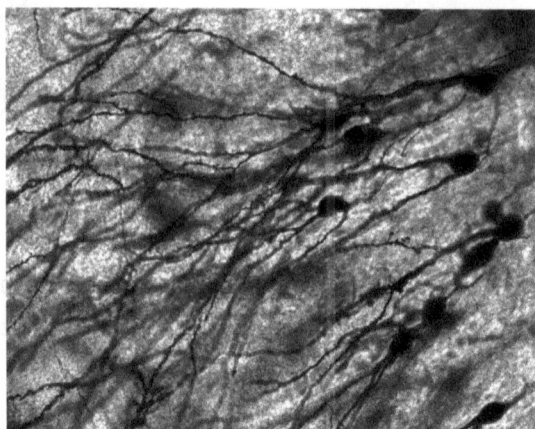

Figure 4-2: Image of neurons in the human hippocampus stained with the Golgi method. 40X magnification. Photomicrograph: ©MethoxyRoxy

Notice how the dendritic extensions of the neuron bodies pack tight together within the interstitial fluid. Dark round and oval spots in the right half of the image delineate neuron bodies. Dendrites appear as long hair-like structures attached to the neuron bodies.

The human brain tissue section shown in *Figure 4-2* was stained using the Golgi silver stain method. Golgi silver staining labels 1%-3% of neurons present in any section, but even so the space appears crowded. In reality, 35-100 times more neurons were in this tissue section than absorbed the silver.

The geometry of the interstitial fluid-filled space around brain neurons modeled by engineers is described as a network of pores and tunnels less than 100 nanometers (10^{-7} meter) across. With an estimated 86 billion neurons in the human brain and ten times that many glia plus 15-25 square meters of blood and lymph vessels, space available to interstitial fluid tunnels is limited. Tight packing creates a setting where chemical communication between neurons and the interstitial fluid occurs across spaces just a few nanometers wide. A nanometer is one part of a meter that is divided into one billion equal parts.

BRAIN'S ELECTRICITY

HOW NEURONS CONNECT

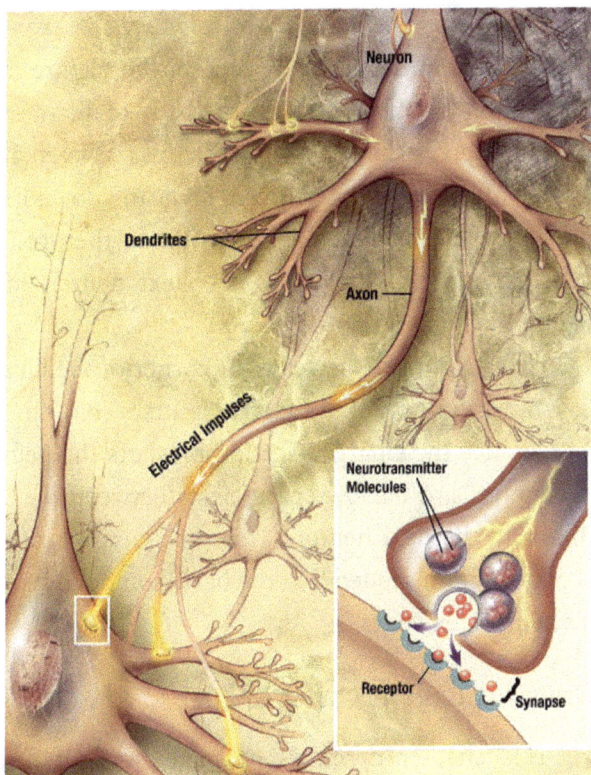

Figure 4-3: How neurons connect with each other. Illustration: This work is in the public domain courtesy of the United States National Institutes of Health.

Deciding where to start when describing how neurons communicate with each other is a challenge, because it is a circular narrative. The receiving end of one neuron responds to input from the messaging end of another neuron. Then, in turn, the receiving neuron sends a message about the signal it received as input to the receiving end of the next neuron (*Figure 4-3*). The dilemma is to decide where in the neuron's structure the story of its electrical membrane should begin.

At the top of *Figure 4-3*, notice where the axon terminals of a neuron outside the frame of the picture make connections on the dendrites and the body of a large neuron. Electric signals shown as jagged arrows proceed along the surface of the receiving neuron's body and down its axon. The axon terminals of the large neuron then connect with the body and dendrites of the next neuron in sequence.

The insert in the lower right corner of the illustration is a magnification of the axon terminal's connection, a synapse. Notice the enlarged section presents the neurons as separate cells. A narrow space exists between the axon terminal and the next neuron. This configuration of neurons connecting in sequence is the basic component of the brain's neuron networks.

ELECTRICAL CURRENT IN THE BRAIN

This chapter describes characteristics of neuron axons (*Figure 4-1*). On occasion axons are compared to electrical wires, because the axon's job is to transmit an electrical signal from the body of the neuron to the axon terminal. However, the electrical wire metaphor is too simple and may be confusing rather than helpful. In reality, while electric currents spread the length of axons, the process is complex and unlike the flow of electrons through a copper wire.

Electrical current in the brain consists of a stream of atoms called *ions*. Ions possess either a positive or a negative charge. The quality and quantity of an ion's charge is governed by the atom's lack of a match between its number of protons, positive particles, and electrons, negative particles.

When the concentration of a particular ion is not the same in the interstitial fluid as in the neuron's cytoplasm, the ion can use open passages in the neuron's membrane to move toward the fluid compartment where its concentration is least. This is a chemical process known as *diffusion*. Diffusion is when molecules relocate by moving through a solution from the place where their concentration is high to another place where their concentration is lower.

The important ions for neuron signaling include sodium ions (Na^+), potassium ions (K^+), chloride ions (Cl^-) and calcium ions (Ca^{++}). The concentration differences of these ions between neuron cytoplasm and interstitial fluid is sufficient to draw them through open tunnel-shaped proteins described as ion channels in the neuron membrane. Ion channel proteins accommodate passage of specific ions. Most ion channels open and close in response to particular signals. In the absence of specific signals, membrane ion channels remain closed, ions cannot diffuse across the membrane and there is no electrical current.

The previous chapter described whole brain mechanisms for keeping the quality of brain's interstitial fluid steady. In addition to strategies for interstitial fluid quality control, all cells including neurons use ion exchange pumps to protect their cytoplasm's ion concentration (*Figure 4-4*).

Ionic Basis of the Resting Membrane Potential

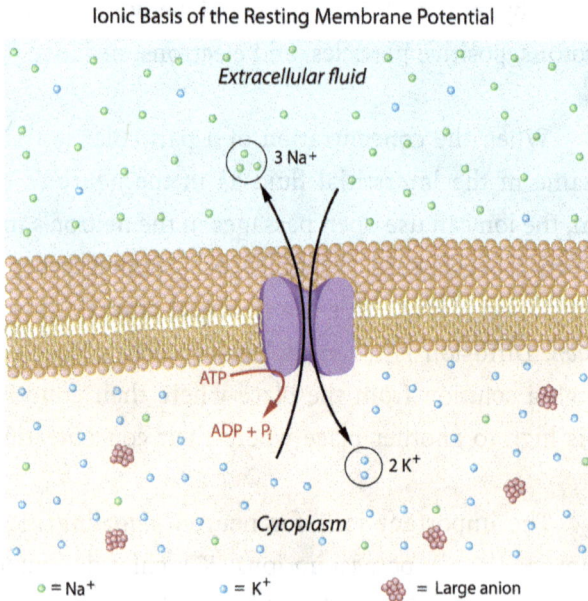

Extracellular fluid

3 Na+

ATP

ADP + P$_i$

2 K+

Cytoplasm

o = Na$^+$ o = K$^+$ = Large anion

Figure 4-4: Energy requiring pumps maintain the ionic composition of neuron cytoplasm. Illustration: ©Alila Medical Images

The pumps are membrane proteins that compensate for ions relocating between cytoplasm and interstitial fluid when ion channels are open. Pumps, moving ions from where their concentration is low to where their concentration is high, require energy from the molecular bonds of adenosine triphosphate, ATP. A great deal of the brain's energy supply is used to power neuron membrane ion pumps.

DIRECTION OF NEURON ION FLOW

When K^+ channels open, K^+ diffuses *OUT* of a neuron's cytoplasm. When Na^+ channels open, Na^+ diffuses *IN-TO* a neuron's cytoplasm. Diffusion caused by ion concentration differences on opposite sides of a membrane is described in physiology as ions moving down their concentration gradient. And, these two ions move down their concentration gradient across neuron membranes creating an opposite flow of positive, chemical-based electrical current.

The opposite flow of Na^+ and K^+ across open channels in neuron membrane appears on the surface to be a simple concept. But the outcome of this simple process is quite powerful in the nervous system. It provides the foundation upon which information is carried by neuronal circuits.

The cytoplasm of neurons, and of most cells, possesses a high concentration of K^+ and of large, negative charged soluble proteins. No membrane channels exist for large molecules so soluble proteins remain in the cytoplasm. They balance the positive charge of the high amount of K^+ within cytoplasm and maintain the cytoplasm's electrical neutrality.

In contrast to K^+, both Na^+ and Cl^- are kept low in cytoplasm of adult neurons. Ca^{++} is high within neurons but stays confined in compartments within the cytoplasm. Ca^{++} is not free to diffuse within cytoplasm except in response to specific triggers. Unlike cytoplasm, interstitial fluid surrounding adult neurons is low in K^+ but high in Ca^{++}, Na^+ and Cl^-. When open channels are available all three ions, Ca^{++}, Na^+ and Cl^- move down their concentration gradient into neuron cytoplasm creating electrical current.

When a neuron is at rest, most ion channels in the axon membrane remain closed and little ionic current can flow. A variety of circumstances cause membrane ion channels of different cell types to open. Each ion channel responds to its own specific opening trigger. Ion channels in a neuron's axon membrane for Na^+ and K^+ open when characteristics of the membrane surrounding them alters in the manner discussed in following sections.

NEURONS AT REST

TRANSMEMBRANE POTENTIALS

All cell membranes have an electrical potential on their inside and their outside surface. An *electrical potential* is an energy created by the presence of electrical charges. When cells are inactive, ions of the cytoplasm and ions of the interstitial fluid create electrical energy of dissimilar magnitude on the two sides of the membrane. A comparison of the electrical potential of the interior side of a cell's membrane to the electrical potential of the exterior side produces a measurable inequality described as a *transmembrane potential*.

Conveniently, the definition of voltage is a difference in electrical potential between two points. When a membrane's potential is different on its two sides, there is a voltage across the membrane. This allows the electrical force difference between the two sides of a cell's membrane to be expressed as a numerical value. Transmembrane potentials of neuron membranes quantify in millivolts, one thousandths of a volt.

Voltage is always a relative term. By convention in physiology the quantity and quality, positive or negative, of transmembrane voltage is always stated as the electrical potential of the inside membrane surface relative to the electrical potential of the outside membrane surface. Therefore, a transmembrane potential of -70 millivolts means the inside surface of the membrane possesses an electrical potential 70 millivolts less than the electrical potential of the outside surface.

The *transmembrane resting potential* is the membrane potential of neurons that are not transmitting signals. The transmembrane resting potential of neurons most often falls in the range of -60 millivolts to -90 millivolts.

ORIGIN OF TRANSMEMBRANE POTENTIALS

Transmembrane potentials exist because a small number of each cell's ion channels for Na^+ and K^+ always remain open allowing a slow constant diffusion of these ions. Ion channels remaining permanently open are referred to as leaky channels or *passive channels.*

Passive ion channels create a transmembrane potential. They differ from the large number of Na^+ and K^+ *voltage-sensitive channels* used by neurons to send a signal. Opening and closing of voltage-sensitive neuron channels is triggered by transient fluctuations of the transmembrane potential and will be described in the next section.

In neuron membrane, passive channels for K^+ outnumber, by far, the passive channels for Na^+. Because K^+ is at a far higher concentration inside the neuron than outside, it diffuses out of the neuron through its passive channels. Incoming Na^+ through the fewer passive Na^+ channels cannot make up for the positive charge lost with the exit of K^+, and the inside of the neuron membrane becomes relatively negative. The magnitude of the negativity of the electrical field on the inside surface of a neuron membrane is determined by the amount of K^+ leaving the cell.

Once outside of the neuron, positive K^+ is drawn to the outer surface of the membrane by the excess of negative charge left at the inside surface. While the exit of K^+ from the neuron through passive channels is driven forward by its high cytoplasmic concentration, other forces stop its exodus. In time the buildup of negative charge on the inside of the membrane holds the remaining K^+ back, and the positive charge built up on the outside of the membrane by the earlier exit of K^+ repels it. By the time forces on K^+ balance, a negative transmembrane resting potential is established.

Little exit of K^+ is needed to create a transmembrane potential. A theoretical transmembrane potential can be calculated for cells with passive K^+ channels but no passive Na^+ channels. This theoretical transmembrane potential is named the *equilibrium potential for potassium*. For neurons, because of the number of passive K^+ channels present, the equilibrium potential for potassium is about -92 millivolts.

The theoretical transmembrane potential when a cell membrane possesses only passive Na^+ channels is the *equilibrium potential for sodium*. For neurons the equilibrium potential for sodium is about +50 millivolts. The inside of the membrane achieves greater positivity than the outside of the membrane by 50 millivolts.

Actual neuron transmembrane resting potentials measure slightly more positive than the -92 millivolts equilibrium potential for potassium. This is because of the contribution of the small number of passive Na^+ channels in neuron membranes.

The final component in the establishment of a stable transmembrane resting potential is the presence of protein ion exchange pumps illustrated above in *Figure 4-4*. Without these energy requiring pumps the passive channels would dissipate ion concentration gradients between the cytoplasm and interstitial fluid. Several kinds of ion exchange pumps exist in the membranes of brain cells in addition to those for Na^+ and K^+.

Ion exchange pumps and ion channels represent different classes of cell membrane proteins with separate spheres of operation. They should not be confused with each other. Ion exchange pumps move ions from one fluid compartment to another against their concentration gradient. Ion channels establish open passages in the membrane allowing ions to diffuse from the fluid compartment where their concentration is high to the fluid compartment where their concentration is low.

VOLTAGE-SENSITIVE ION CHANNELS

ION CHANNEL STRUCTURE

Characteristics of protein molecules make them excellent ion channel structures. Proteins consist of assemblies of 20 unique molecular units named amino acids. Individual amino acids are fat-soluble or water-soluble. Some are large molecules and some are small. Some amino acids carry a charge, and some remain neutral.

To build a protein, amino acids connect together in an arrangement described as a peptide bond. The amino acids link one after the other to form a chain named a peptide because of nature of the bonds. Some protein chains become long, greater than 2000 amino acids, and some remain short, 10 to 200 amino acids.

Protein chains fold over on themselves because of the different size and charge of individual amino acids. Each amino acid searches for space and a comfortable electrical field environment. The portions with positive charge repel each other. The portions with negative charge repel each other, and the opposite charged portions gather together. One configuration, favored by some sections of proteins as they fold, is a spiral shape named an alpha helix, a right-handed coil (*Figure 4-5*).

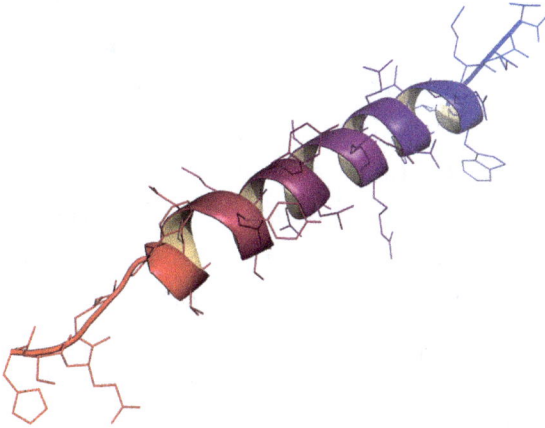

Figure 4-5: Section of a protein showing open-loop peptide backbone structures on both ends with an alpha helical peptide backbone structure between. The line drawings attached to the backbone structures represent amino acid side chains. Amino acid side chains distinguish one amino acid from another. Illustration: ©molekuul.be

If the sequence of amino acids in a helix is arranged where all the neutral amino acid side chains occur on one side of the helix and the charged amino acid side chains occupy the opposite side, the helix is able to insert itself in a cell's lipid membrane with the neutral amino acid side chains becoming membrane anchors. The part of the helix away from the membrane is able to retain electrical properties attractive to water and ions. Protein sections passing through cell membranes always have an alpha helical form.

Another feature of proteins, which makes them exceptional molecules for constructing ion channels, is their ability to adjust their shape when other molecules enter their electrical field. When the transmembrane resting potential of a neuron deviates from its normal value, voltage-sensitive protein channels respond by changing their shape. The resulting change in the position of the amino acid side chains opens a tunnel through the membrane.

Voltage-sensitive channel proteins for Na^+ and for K^+ in neuron axon membranes appear similar to each other in structure. The basic organization of a voltage-sensitive ion channel consists of four domains each with six transmembrane alpha helices.

The voltage-sensitive K^+ ion channel (*Figure 4-6*) is a composite of four separate proteins, each with a configuration of a single domain with six transmembrane alpha helices. In contrast the four domain arrangement is shown spread out as a single amino acid chain for the voltage-sensitive Na^+ channel in *Figure 4-7*.

Exclusivity of K^+ Channels

Membrane ion channels allow exclusive passage of particular ions. How ion channel selectivity is accomplished remained a mystery until recent data uncovered the origin of the uniqueness of the ion channels for K^+ and Na^+. These data solve the puzzle of why Na^+, a smaller ion, cannot pass through ion channels similar in structure designed for the larger K$^+$. The K$^+$ channel diagramed in *Figure* 4-6 presents an excellent example of how a protein channel is organized to favor a particular ion.

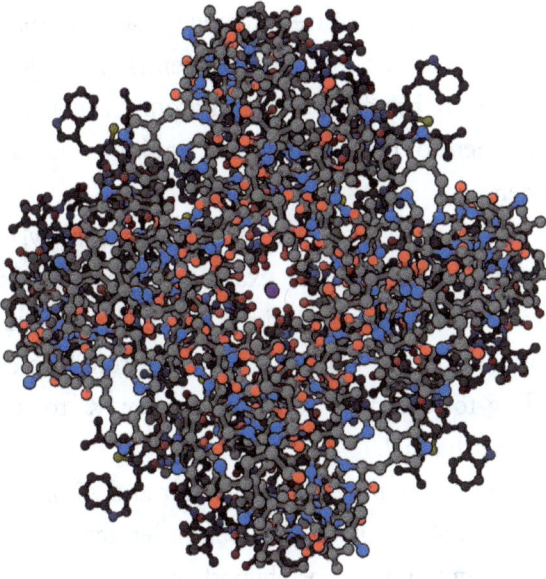

Figure 4-6: View of potassium (K$^+$) in the filter of a K$^+$ channel. This view is from inside the cell. K$^+$ is the purple dot at the center of the structure. The red dots represent oxygen atoms; gray dots denote carbon atoms; blue dots signify hydrogen atoms. Notice the symmetrical placement of four identical subunits. Illustration: Based upon Public Data Bank 1BL8 is released to public domain by Bensaccount at Wikimedia project.

When ions like Na$^+$ and K$^+$ disperse in water they become surrounded by, and bound to, water molecules by a type of weak chemical bond described as a hydrogen bond. In order for ions to pass through the K$^+$ channel's filter, the sphere of hydrogen bonded water molecules surrounding K$^+$ must be stripped away. Stripping off molecules of water requires a hydrated ion to encounter within the ion channel other atoms capable of displacing the water molecules.

Structural analysis of voltage-sensitive K^+ channels, similar to those found in the human brain, revealed a passage with a narrow filter area created by carbonyl oxygen in four critical positions. Carbonyl oxygen is an oxygen atom with two strong bonds to a carbon atom. Carbonyl oxygen is a component of the peptide bond linking amino acids to each other.

In this particular channel, the four carbonyl oxygen components localize to the exact positions necessary to drive water away from around the K^+. Yet, because the oxygen atom of carbonyl oxygen is bonded strongly to carbon, its binding to K^+ is weak enough to allow K^+ to continue through the channel.

The precise movement of water and K^+ through the voltage-sensitive K^+ channel is still under active investigation. A contemporary theory hypothesizes one water molecule from each hydration sphere also passes through the channel's filter. The sequence of molecular movement through the selectivity filter is thought to be alternating potassium ion and water molecule, K^+-H_2O-K^+-H_2O-K^+-H_2O.

The same process does not work for Na^+ at the K^+ channel, because Na^+ is smaller than K^+. The four carbonyl oxygen components forming the selectivity filter of the K^+ channel are spaced too far apart to rid the smaller Na^+ of its hydrogen bonded water molecules. With its water hydration sphere left intact, Na^+ is too large to pass through the K^+ channel.

SELECTIVENESS OF NA⁺ CHANNELS

Voltage-sensitive Na⁺ channels located in neuron axons consist of one large alpha protein associated with two or three smaller beta proteins. The alpha protein contains the channel with a selective filter for Na⁺. The beta proteins alter the cellular location of the alpha channel and alter its voltage sensitivity.

Figure 4-7 shows the alpha subunit of a Na⁺ channel spread out in a membrane to illustrate it in detail.

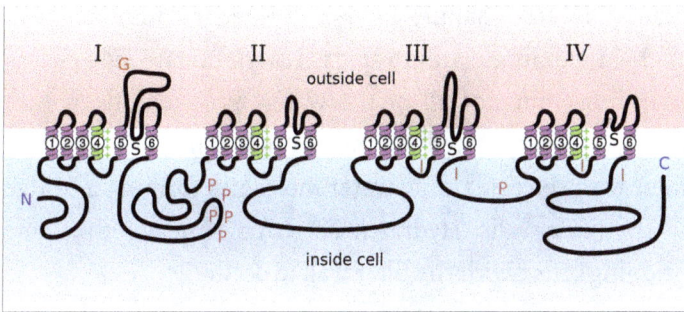

Figure 4-7: Sections of the alpha subunit of a voltage-sensitive Na⁺ channel spread out in a membrane (white band). Notice the four domains, I, II, III and IV, each with 6 transmembrane alpha helices through the membrane. G = sites where sugars can be attached, P = amino acids where phosphate groups can be attached, S labels the location of the ion channel filter components, I = amino acids affected when the channel closes after being open. Illustration: ©Cthuljew

In the axon membrane, the four domains labeled I-IV form a circular cluster resembling the four subunits of the K⁺ channel shown in *Figure 4-6*. When the domains cluster, the protein sections illustrated as loops outside the cell close over the areas labeled S in the diagram. These loops create the Na^+ channel filter. The remainder of the channel in the membrane is formed by the four sets of helices numbered 5 and 6 in the illustration. The area between helices 5 and 6 becomes the open channel linking the interstitial fluid with the neuron's cytoplasm.

Even though the overall organization of the alpha subunit of the voltage-sensitive Na^+ channel is similar to the voltage-sensitive K^+ channel, the contour of its ion selectivity filter is quite different. The entrance of the Na^+ channel filter is larger than the entrance of the K^+ channel filter. The Na^+ channel filter is formed by four negative charged glutamate amino acids. The glutamate side chains strip most but not all of the water molecules from Na^+. Small, partly hydrated Na^+ moves through its ion channel filter with the remaining water molecules forming a bridge to the filter walls. Hydrated K^+ is too large for the same bridging effect with the filter wall to occur.

Axon Signaling

Arrangement of Voltage-sensitive Channels

The position of the voltage-sensitive Na^+ and K^+ channels along the length of the axon is critical to fidelity of electrical conduction along neuron axons. The well-ordered clusters of voltage-sensitive Na^+ and K^+ channels, and their sequential pattern of opening and closing, transmit chemical electricity from the axon's initiation site near the axon hillock of the neuron body to the terminal end of the axon.

To paint a clear picture of how clusters of Na^+ and K^+ voltage-sensitive channels distribute along axons another type of brain cell, the oligodendrocytes, must be introduced. Oligodendrocytes possess a cell body central to many long membrane extensions (*Figure 4-8*). Oligodendrocytes use their slender extensions to hang on to axons.

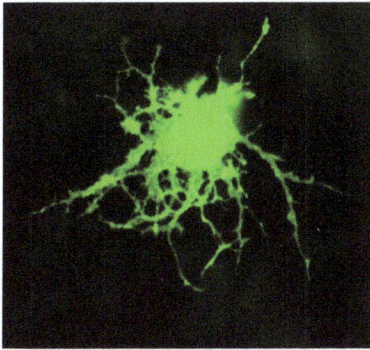

Figure 4-8: A mouse oligodendrocyte labeled with GFP (Green Fluorescent Protein). The image was recorded using a fluorescent microscope with a 63x objective. Photomicrograph: ©Jurjen Broeke

Where oligodendrocyte extensions contact neuron axons they convert to flat sheets. The flattened sheet of membrane winds around neuron axons creating a cover with many layers. The layered portion of the oligodendrocyte membrane compacts and pushes out most of its own cytoplasm shrinking the wrap to tight layers of membrane.

The membrane of oligodendrocyte extensions contains a high level of a variety of fat molecules. The entire group of fat molecules is described collectively as *myelin*. Myelin acts as an axon's electrical insulation. Myelin also protects fragile axons against inflammatory and oxidative injury. A single oligodendrocyte may enclose a single axon or it may reach out and enclose the axons of many neurons at once.

For simplicity, illustrations showing myelination of neurons leave out oligodendrocyte cell bodies and picture just the covers made by individual membrane extensions as shown in orange in *Figure 4-9*. The layer of myelin is not continuous along the axon. Between the myelin layers bare patches of axon occur. The bare patches, known as Nodes of Ranvier, are named for the man who discovered them, French pathologist and anatomist Louis Antoine Ranvier.

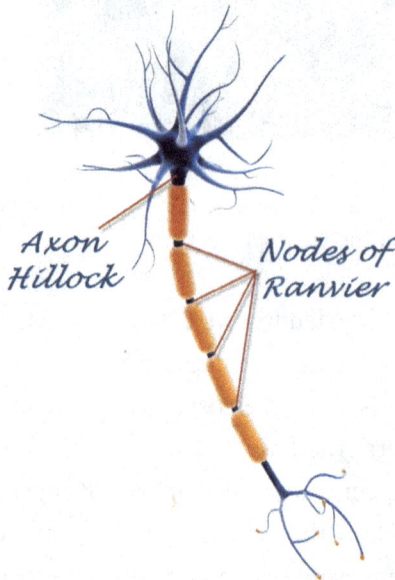

Figure 4-9: Drawing of a neuron showing myelin sheaths colored orange. The oligodendrocytes cell bodies are not present. Illustration: ©martan

Dense clusters of voltage-sensitive Na^+ channels occur at the Nodes of Ranvier. The assembly of Na^+ channels at the Nodes of Ranvier includes secretion of adhesive proteins by the oligodendrocytes. Clusters of the channels form around the adhesive proteins. Fibrous protein components within the axon, collectively described as its cytoskeleton, also play a role in anchoring Na^+ channels in place.

In contrast to the position of the voltage-sensitive Na^+ channels on the axon, voltage-sensitive K^+ channels gather together under the edges of the myelin wrap. The edges of the oligodendrocyte wrap merge with the axon membrane to form a thin partition separating Na^+ and K^+ channel clusters. Axonal K^+ channels reside near, but inside, the partitions under the myelin.

Axon myelination is an ongoing process throughout brain development and into adulthood. The frontal lobe of the cerebral cortex, the area associated with learning new concepts and memory, is the last brain region to add myelin to its axons. It is also the area in adults with the maximum amount of axon myelination.

The density of axons with myelin in the adult human brain varies to a great extent among regions. Many axons in the brain lack oligodendrocytes. Axons without myelination also display voltage-sensitive Na^+ and voltage-sensitive K^+ channels in alternating clusters spread along the axonal membrane but the speed of signal transmission is comparatively slow. Axons with myelin conduct signals about 50 times faster than axons without myelin.

Oligodendrocytes may influence the design and function of brain circuitry. Patches of myelin on living axons appear in an irregular pattern, unlike myelin illustrated in *Figure 4-9*. Some neurons in the cerebral cortex maintain extensive stretches of axon without myelin, followed by stretches with myelin. Many theories exist about why this is. One speculation is larger Nodes of Ranvier allow neighboring neurons to synapse on the bare parts of an axon. Another guess is myelin sheaths produce chemicals to block axon collateral formation. Axons split and send collaterals off to synapse on multiple target cells, but axon collaterals only form where an axon is bare. Both synapses on axons and axon collaterals participate as central elements of complex brain circuits.

TRANSIENT MEMBRANE VOLTAGE PATTERNS

The signal traveling along an axon is electrical because ions flowing through Na^+ and K^+ voltage-sensitive membrane channels carry a charge. Moving a signal along an axon requires setting in motion a pattern of voltage transients, short-lived and rapid fluctuations in the transmembrane potential called *action potentials* (*Figure 4-10*). Consecutive action potentials progress the length of an axon producing a signal that maintains its strength as it travels from its initiation site to the axon's terminal end.

Voltage-sensitive Na^+ and K^+ channel filters remain closed to flow of ions at the axon's negative transmembrane resting potential. When a transmembrane potential moves away from its negative resting value to a sufficiently less negative voltage, a threshold potential, voltage-sensitive ion channel filters open for a short period of time.

A positive drift of the transmembrane resting potential at the initial segment of the axon is the result of events within the neuron body at the axon hillock (*Figure 4-9*). Diffusion of Na^+ from the axon hillock depolarizes the initial segment of an axon membrane to its threshold potential.

Depolarize means to bring electrical potentials on two sides of a membrane closer to being equal. The threshold potential causes the first cluster of the axon's voltage-sensitive Na^+ channels to open (*Figure 4-10*).

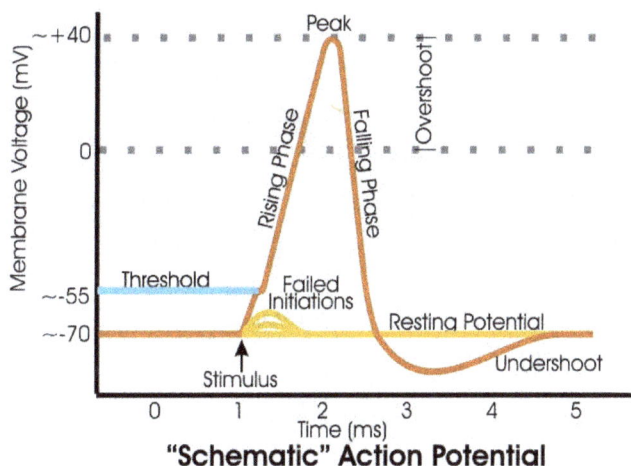

"Schematic" Action Potential

Figure 4-10: Transmembrane voltage transients during an action potential. Notice the time between the membrane reaching threshold and the peak change in transmembrane voltage is less than one thousandth of a second (ms). The magnitude of the threshold potential needed for initiation of an action potential varies among neurons. Illustration: ©Synaptidude

The axon hillock is also able to originate action potentials, but its threshold potential is more positive than the axon's threshold potential. Membrane depolarization at the axon hillock is a result of action potential–like activity named dendritic spikes spreading over the membrane of the neuron body from the dendrites. Dendritic spikes, the result of signaling at neuron synapses, are explained further in Chapter 5, *"Neuron Synapses—Excitatory and Inhibitory."*

The electrical field adjustment associated with depolarization of the transmembrane potential changes the alignment of amino acid side chains in the Na^+ channel protein. Repositioning of amino acid side chains opens the Na^+ channel's ion filter and permits passage of Na^+ across the membrane into the axon's initial segment. Na^+ diffusing through open channel filters, in turn, moves the transmembrane potential nearer to zero millivolts. A representative action potential is diagramed in *Figure 4-10*.

SPREAD OF ACTION POTENTIALS

The transmembrane threshold potential necessary to open an axon's voltage-sensitive Na^+ channels is reached before the threshold potential required for opening the K^+ channels. That is, a membrane depolarization of greater magnitude is required to open voltage-sensitive K^+ channels. Because of this difference in response to changes in the transmembrane potential, axon Na^+ channels always open before K^+ channels.

Open voltage-sensitive Na^+ channels allow Na^+ to diffuse into the axon until it approaches the equilibrium potential for sodium—the potential where the positive repelling environment within the axon cytoplasm overcomes the force of the concentration gradient driving Na^+ into the cell. The peak transmembrane potential of an action potential is a bit less than the actual Na^+ equilibrium potential for two reasons. First the voltage-sensitive K^+ channels begin to open before the potential peaks allowing K^+ to carry positive charge out of the axon.

Second, the entire cluster of Na^+ channels is not open at the same time. Before the peak of the action potential, part of the group of Na^+ channels begins a gradual close. Refer back to *Figure 4-7* and notice the Na^+ channel filter is located at a different place on the alpha channel protein than the inactivation sites that cause it to close. This arrangement of the Na^+ channel's voltage-sensitive amino acids makes opening and closing the Na^+ channel separate processes. Once the channel filter closes, a period of time must pass before the protein rearranges itself into a position where the filter area can again respond to a transmembrane potential.

As Na^+ enters the axon, it diffuses away from its channel. It spreads over the inside of the membrane in all directions. Some of the Na^+ travels toward the axon terminal under the myelin wrap in the direction of the next Node of Ranvier. When sufficient Na^+ reaches the next Node of Ranvier, it depolarizes that segment of the axon to threshold and elicits another action potential. This process repeats itself from node to node until an action potential is produced at the axon terminal.

Some of the Na^+ entering the axon also diffuses back toward the neuron's body. At the axon hillock, back diffusion of Na^+ brings the transmembrane potential of the axon hillock to threshold producing an action potential that spreads along the membrane of the neuron body. The body's action potential travels toward the dendrites. How far this action potential moves over the membrane surface and into the dendrites depends upon the membrane's pattern of voltage-sensitive ion channels. The distribution of these channels varies from neuron to neuron. The flow of membrane action potentials from the axon hillock toward the dendrites informs the rest of the neuron that an output signal was generated.

The sequence of ion channel events along an axon insures axonal action potentials move along in one direction, from the initial segment to the axon terminal. Any Na^+ diffusing in the direction of ion channels at the previous node is unable to bring the transmembrane potential at the previous node to threshold. This is because the later opening K^+ channels remain open there for K^+ to diffuse out of the axon carrying positive charge with it.

Not all the ion channels in a cluster open and close in unison. There is a time during each action potential when most of the Na^+ and K^+ channels are open. During this lull before channels close, the axon cannot produce another action potential. This pause is named the axon's *absolute refractory period*. An axon's absolute refractory period limits the transmission frequency of action potentials and helps to force action potentials to progress toward the axon terminal. An absolute refractory period lasts about one thousandth of a second.

Some axons also exhibit a *relative refractory period* when most, but not all, of the voltage-sensitive Na^+ and K^+ channels again respond to depolarization of the transmembrane potential. During the relative refractory period a greater than normal membrane depolarization is required to elicit an action potential. This is because the transmembrane potential immediately following an action potential is closer to the potassium equilibrium potential than to the normal transmembrane resting potential. Relative refractory periods last about 2 thousandths of a second.

AXON HOUSEKEEPING

Components within neuron cytoplasm also transfer along axons. Axon terminals depend upon the machinery of the neuron body for synthesis of structural elements. For an action potential to create the desired effect at the axon's terminal end, the terminal end must possess an adequate stock of proteins, mitochondria, vesicles, nutrients, energy molecules and other critical items. Everything required for well-being of the axon terminal must be transported from the neuron body through the axon.

A feedback messaging system is also necessary to regulate appropriate timing of the flow of molecules and items from the neuron body to the axon terminal. This means substances must be transported in both directions.

Two terms describe the direction of flow of biologic material in axons. Items transported *anterograde* move toward the axon terminal. Items transported *retrograde* travel toward the neuron body from the direction of the axon terminal.

In neurons, as in other cells, shuttling of cytoplasmic components from place to place is a continuous process requiring energy. Re-arrangement of biologic molecules and structures is accomplished by motor proteins binding cargo and conveying it along fibrous tracks described as *cytoskeleton*.

Axons contain an extensive cytoskeleton along their entire inner length. It provides a scaffold for cargo delivery to and from the neuron body. Things travelling in membranous sacks move with speed along the axon's cytoskeleton. Enclosed material includes among other things, amino acids, various proteins, lipids and sugars.

Motor proteins bind to specific membrane proteins on the sacks either directly or with the help of cytoplasmic linker proteins. Cargo not enclosed in membranous sacks moves sluggishly through the length of the axon. The mechanism for movement of unpackaged items along the cytoskeleton also includes motor proteins, but the process is less continuous. Slow moving items progress partway along the axon and then stop for a period of time and then move on again. The intermittent nature of this process suggests there may be multiple monitoring points along the way.

SUMMARY CHAPTER 4

- Ions, charged atoms, carry the chemical-based electrical currents of the brain
- The difference in the concentration of ions between neuron cytoplasm and brain interstitial fluid permit ions to diffuse through open membrane channels
- Sodium ion (Na^+), potassium ion (K^+), chloride ion (Cl^-) and calcium ion (Ca^{++}) create the electrical currents in brain
- Neuron membranes display a different electrical potential on the side bathed by interstitial fluid than on the side bathed by cytoplasm
- Transmembrane voltage, which is measured in millivolts, is always stated as inside electrical potential relative to outside electrical potential
- A transmembrane potential of -70 millivolts means the inside surface of the cell membrane possesses an electrical field force 70 millivolts less than the electrical field force of the outside surface
- Voltage-sensitive ion channels of the axon membrane occur as large protein complexes
- Voltage-sensitive channels open when the transmembrane potential near them reaches a threshold value
- The transient fluctuation of the transmembrane potential of neuron axons, caused by the opening and closing of voltage-sensitive Na^+

and K$^+$ channels, is known as an action potential

- Action potentials travel along an axon from the neuron's body to the axon terminal
- Materials move between the axon's terminal and the neuron's body along the axon's cytoskeleton

[5]

Neuron Synapses—
Excitatory and Inhibitory

ARRIVAL OF ACTION POTENTIALS at axon terminals sets in motion events supporting communication between neurons. Transfer of information from neuron to neuron, or from neuron to non-neuron cell, is in most cases a molecular rather than electrical phenomenon.

Santiago Ramón y Cajal shared the 1906 Nobel Prize with Camillo Golgi. They received the prize for their extensive body of work describing the structure of brain neurons. Santiago Ramón y Cajal proposed neurons may be separate entities that communicate with each other across narrow spaces of about 20 nm (20 x 10^{-9} meter). Neuroscientists did not embrace this idea at first. However, Dr. Ramón y Cajal was correct. The structures at the places where neurons connect with each other are now collectively assigned the name *synapse* (*Figure 5-1*). Synapses and their surrounding structures attract considerable ongoing attention.

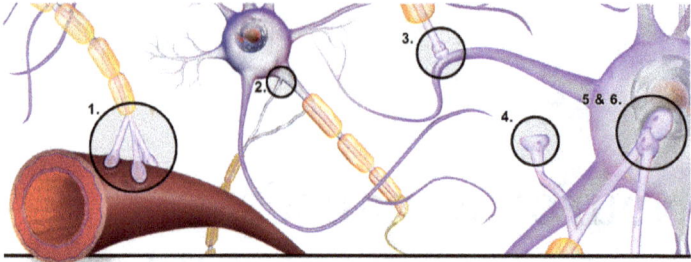

Figure 5-1: Some locations of neuron synapses found in the body. Illustration: ©BruceBlas

BRAIN SYNAPSES

Brain synapses occur at neuron dendrites, at the neuron cell body and at axons. But, neurons throughout the body, peripheral neurons, also form synapses on other cells like skeletal muscle and blood capillary smooth muscle.

The complex placement of synaptic structures in the brain permits fine-tuned regulatory control of neuron input and output and creates complex neuronal circuits that act as units.

The most common type of synapse in the brain is one where a small molecular weight chemical called a *neurotransmitter* is released from the axon terminal. Neurotransmitter is stored in the axon terminal in vesicles. The section of the axon terminal membrane dedicated to neurotransmitter release is called the *presynaptic compartment* (*Figure 5-2*) of the synapse.

Neurotransmitter released from the presynaptic compartment diffuses into a gap of about 20 nanometers between the presynaptic compartment and a modified section of neuron cell membrane across the gap named the *postsynaptic density (Figure 5-1)*.

The gap itself is called a *synaptic cleft*. When the postsynaptic cell is a brain neuron, the postsynaptic density may be on a dendrite, on a dendritic spine, on the neuron body, on the axon or on an axon's terminal.

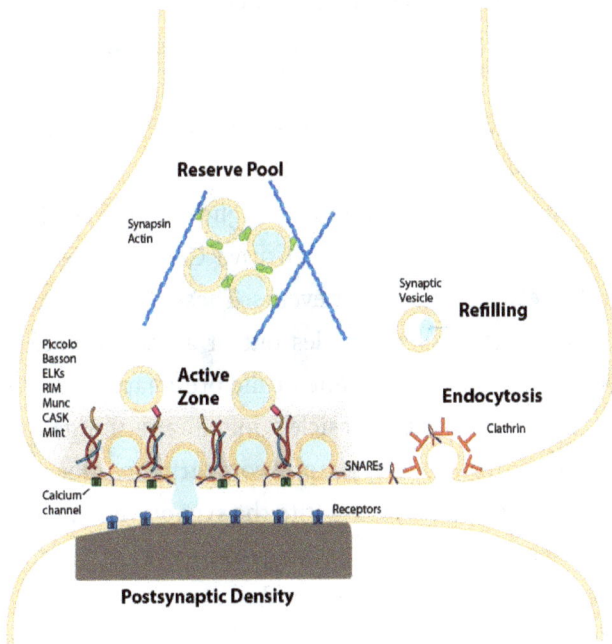

Figure 5-2: Illustration of some of the key elements found at neuron to neuron synapses. Illustration: ©Curtis Neveu

Presynaptic Compartment

Neurotransmitter Release

Neurotransmitter is stored by the presynaptic neuron in vesicles inside axon terminals (*Figure 5-2*). When action potentials arrive at the axon terminals, the membrane depolarization they initiate causes voltage-sensitive calcium ion (Ca^{++}) channels to open. Ca^{++} diffuses through the open channels into the presynaptic compartment from the interstitial fluid. Ca^{++} entering the neuron is called free Ca^{++} to distinguish it from cytoplasmic Ca^{++} bound to cellular proteins.

Free Ca^{++} is a thousand times higher in the interstitial fluid surrounding adult brain neurons than in neuron cytoplasm. The introduction of a high level of free Ca^{++} to the cytoplasm when voltage-sensitive Ca^{++} channels open sets off a series of molecular events. These events transport neurotransmitter filled vesicles out of a reserve pool and deliver them to the membrane of the presynaptic compartment. Upon arrival, the vesicle's membrane merges with the membrane of the presynaptic compartment, opens and releases its neurotransmitter into the synaptic cleft.

The axon terminal's voltage-sensitive Ca^{++} channels contain one of five different combinations of four protein subunits. All voltage-sensitive Ca^{++} channel subunit combinations include an alpha subunit with the ion channel. The Ca^{++} alpha subunit displays an organizational pattern similar to the axon's Na^+ channel alpha subunit described in the previous chapter. The alpha subunit of the voltage-sensitive Ca^{++} channel contains a filter specific for Ca^{++}.

The alpha subunits of voltage-sensitive Ca^{++} channels vary in their amino acid composition depending upon their location in the body and their precise physiologic function. Three variations of the Ca^{++} alpha subunit reside in the brain to service neurotransmitter release. Three associate proteins help anchor it in the axon terminal membrane and modulate its filter properties, affecting its activation threshold and rate of inactivation. Associate proteins also control transfer of the Ca^{++} alpha subunit protein from its place of synthesis in the neuron to the axon terminal membrane.

RETRIEVAL OF NEUROTRANSMITTER

Neurotransmitter wears out its welcome in a synaptic cleft in a short period of time. Action potentials travel to an axon terminal at a fast pace, often just a few thousandths of a second apart. If neurotransmitter released by each signal is not removed rapidly, the synaptic cleft becomes saturated making further signals ineffective.

Most neurons depend upon neurotransmitter reuptake pumps called *transporter proteins* to remove it from the synaptic cleft, and thereby end its effectiveness at the next neuron. Neurotransmitter transporter proteins represent one of the multiple classes of proteins found in the membrane of the presynaptic compartment. In many areas of the brain, neurotransmitter in the synaptic cleft is also transported into astrocyte glial cells surrounding neuron synapses. Over twenty recognized proteins comprise the collection of brain neurotransmitter transporters. Each neurotransmitter uses a special subset of the transporter proteins.

Transporter proteins take advantage of ion concentration differences on the two sides of synaptic membranes to supply the required energy for pumping neurotransmitter out of the synaptic cleft. These transporters also go by the generic name of *symporters*. Symporters move two or more different molecules, or ions, in the same direction across cell membranes with at least one of them diffusing down its concentration gradient and one being moved against its concentration gradient. At the synapse, the molecule moving against its concentration gradient is the neurotransmitter. In fact, the brain uses much of its energy pumping symporter ions back out of the cell.

Most neurotransmitter symporters depend upon Na^+ and Cl^-. Interstitial fluid in the synaptic cleft is higher in Na^+ and Cl^- than the neuron cytoplasm. The transporter protein's shape allows Na^+ and Cl^- to diffuse across the neuron membrane into the cytoplasm and carry neurotransmitter with them. Ion concentration gradients can be thought of as indirect sources of energy, because energy consuming pumps are required to maintain them.

An exception to reuptake of neurotransmitter from the synaptic cleft by transporters occurs when an axon terminal releases the neurotransmitter named acetylcholine. Acetylcholine is inactivated in the synaptic cleft rather than transported back into the presynaptic terminal. There the enzyme acetylcholinesterase splits it into choline and acetate.

Acetylcholinesterase is located on the interstitial fluid side of the postsynaptic membrane. Choline is then transported back into the presynaptic compartment for synthesis once again into acetylcholine. This may seem like a complicated approach. But, the result is a rapid rate of removal of acetylcholine from the synaptic cleft. The achieved rate would be unattainable using a transporter.

POSTSYNAPTIC COMPARTMENT

POSTSYNAPTIC DENSITY

Synapses on neuron dendrites continue to be the brain synapses most often studied. The postsynaptic compartments of neuron dendrites occur on small projections of the dendrite's membrane and on the dendrite's main shaft.

The small projections are named *dendritic spines* because of their shape. Neurons with extensive, branched dendrites may contain up to thousands of dendritic spines per branch (*Figure 5-3*). The greater the branching of its dendrites, the larger the volume of incoming signals a neuron accommodates.

Dendrites experience simultaneous neurotransmitter input from multiple presynaptic neurons. The response of the dendrite depends upon its complement of proteins capable of interacting with neurotransmitters. The location and distribution of various types of neurotransmitter-response proteins influence the net signal passed on to the neuron cell body. Postsynaptic proteins that respond to neurotransmitter are called *receptors*.

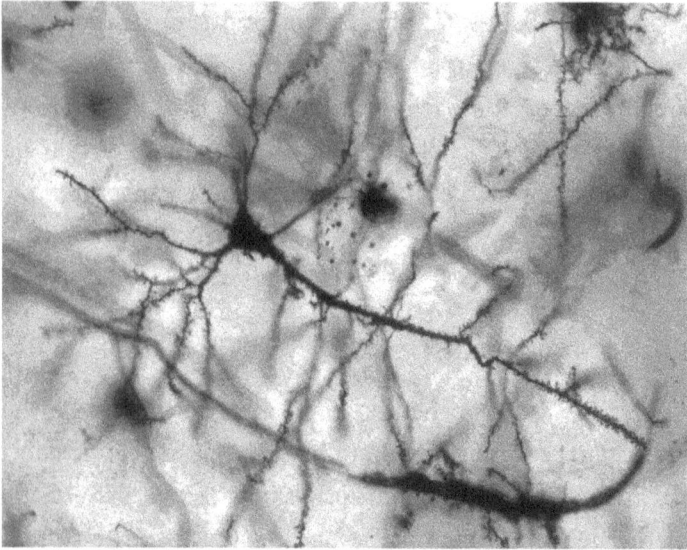

Figure 5-3: A pyramidal neuron in a human hippocampus stained with the Golgi method at 40X magnification. Notice the many small projections on the surface of the long dendrites extending from the cell body, the black triangular formation. Photomicrograph: ©MethoxyRoxy

A dominant feature of postsynaptic compartments is a structure called the *postsynaptic density* (*Figure 5-2*). A postsynaptic density is a thickening of the postsynaptic membrane. It contains an extensive collection of proteins that move about in the membrane, including neurotransmitter receptors. The protein population of postsynaptic densities fluctuates based upon the amount and type of incoming neurotransmitter from presynaptic neurons.

Postsynaptic density proteins fall into several classes including adhesion molecules, enzymes and receptors. Adhesion molecules hold the two neuron membranes together stabilizing the synapse. The purpose of the numerous postsynaptic density enzymes is still open to speculation. The best understood brain synapse enzymes cut a small protein, amyloid-β, from its large membrane bound precursor protein. Amyloid-β physiology is discussed in detail in Chapter 8, *"When It All Goes Wrong— Alzheimer's Dementia."*

Receptor is a generic term describing a broad class of proteins activated by specific chemicals called *ligands*. In the brain, neurotransmitter is a ligand for the postsynaptic density receptors. Receptors act through a variety of mechanisms to regulate cell performance. In general, receptor proteins respond to ligands by setting in motion a precise sequence of cellular events. The cellular activities influenced by a particular receptor depend upon the structure of the receptor, its ligand and its cellular location.

About a dozen different neurotransmitters act as receptor ligands to trigger brain's postsynaptic receptors. Two neurotransmitters, the biologic molecules called *glutamate* and *γ-aminobutyric acid (GABA)*, control most of the brain's neuron activity when a person is awake. The postsynaptic effect of glutamate on membrane ion channels is estimated to consume about one third of the entire energy used by the brain. Most of brain's energy consumption is by ATP-requiring ion exchange pumps that maintain the transmembrane potential.

Ligands glutamate and GABA trigger different dendrite receptors and open different ion channels. Open glutamate receptor channels permit entry of Na^+ and Ca^{++} into dendritic spines. Open GABA receptor channels permit Cl^- to enter, and K^+ to leave, the dendrite's shaft. Entry of Na^+ and Ca^{++} through glutamate receptors depolarizes the neuron membrane making the transmembrane potential more positive. Entry of Cl^- and exit of K^+ through GABA receptors generates an extremely negative, hyperpolarized transmembrane potential.

When the transmembrane potential is hyperpolarized, the Na^+/Ca^{++} receptor channels cannot open, even in the presence of glutamate. Thus, GABA receptor activation blocks the purpose of glutamate receptors. Glutamate is called an *excitatory neurotransmitter* because it causes action potential-like dendritic spikes. GABA is called an *inhibitory neurotransmitter* because it blocks the ability of glutamate to initiate dendritic spikes.

NEUROPLASTICITY

The process by which the size and shape of dendritic spines alter over time, a progression called *neuroplasticity*, is one of the most exciting areas of contemporary brain research. Dynamic dendritic spines probably form the anatomic infrastructure of memory.

Neuroscientists believe that neurons with a great number of large spines on their dendrites are most likely to participate in formation of complex memories. Dendritic spines experiencing abundant input from presynaptic neurons persist and enlarge. Those receiving few signals shrink to a stubby shape and sometimes disappear completely from the dendrite. Spine shapes usually observed are shown in *Figure 5-4*.

Spine types

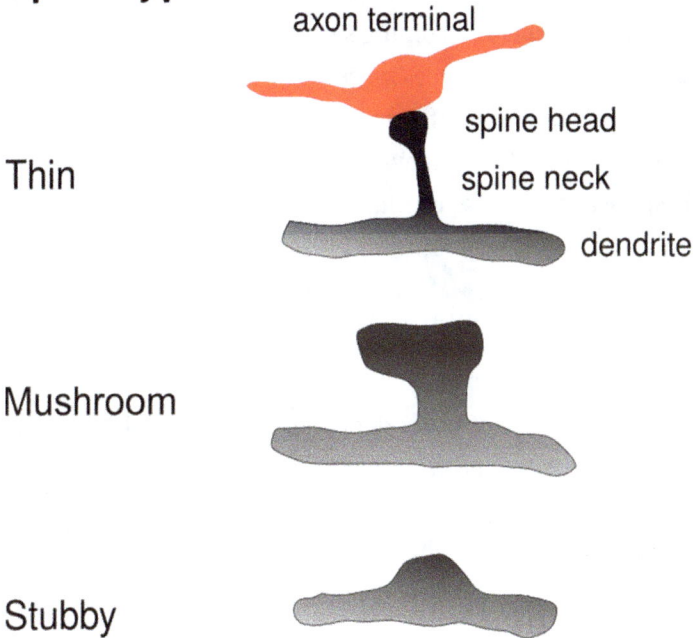

axon terminal

spine head

Thin

spine neck

dendrite

Mushroom

Stubby

Figure 5-4: Illustration depicting main shapes for dendritic spines. This illustration is released to public domain by the artist

Dendritic spines operate as discrete neuron compartments. They display a high degree of molecular organization. Postsynaptic glutamate receptor activation produces two general categories of action within dendritic spines, an electrical action and a biochemical action

Opening of glutamate receptor ion channels permits entry into the dendritic spine of both Na^+ and Ca^{++}. Entering Na^+ flows along the membrane of the dendritic spine into the dendrite. Na^+ depolarizes the dendrite's membrane toward the threshold needed for initiation of dendritic spikes, which are similar to action potentials, to alert the neuron of incoming signals.

In contrast, free Ca^{++} within the dendritic spine sets in motion a series of intracellular biochemical reactions that ultimately create new molecules to enlarge the dendritic spine and increase its number of neurotransmitter receptors.

ELECTRICAL OUTPUT OF DENDRITIC SPINES

Na^+ current, originating from many dendritic spines is summed in the dendrite. The summed Na^+ initiates a dendritic spike. Although dendritic spikes resemble action potentials of the neuron's axon, the location of the voltage-sensitive ion channels of the dendrite lack the rigid organization of an axon's voltage-sensitive ion channels. Information about organization of axon voltage-sensitive ion channels is presented in Chapter 4, *"Neurons—How They Make Electricity."*

Action potential transmembrane transients of axons are achieved using only Na^+ and K^+ voltage-sensitive channels. In contrast, dendritic spikes require Na^+, Ca^{++} and K^+ voltage-sensitive channels. Dendritic Na^+, K^+ and Ca^{++} voltage-sensitive channels open when the transmembrane potential of the dendrite reaches an appropriate threshold value. Voltage-sensitive K^+ channels always open last. The threshold potential for a dendritic spike is far higher than the threshold for an action potential at the axon. Therefore, a greater increase in Na^+ in the cytoplasm is required for initiation of a dendritic spike.

Dendritic spikes move toward the neuron cell body slower than action potentials move along an axon. The rate of dendritic spike progress depends upon the number and pattern of voltage-sensitive channels present and on the degree of branching of the dendrite before the signal arrives at the cell body. Near the cell body dendritic spikes are stronger than dendritic spikes originating at distal points on the dendrite. Dendritic spikes flow over the neuron cell body membrane and merge with the cell body action potentials initiated by backflow of Na^+ from action potentials of the axon. The net effect of the electrical potential along the cell body membrane keeps the metabolic machinery of the neuron aware of both incoming and outgoing signals.

BIOCHEMICAL RESPONSE TO NEUROTRANSMITTER

New biochemical molecules created in response to the entry of free Ca^{++} into a dendritic spine, transfer through the spine neck and the dendrite to the nucleus in the neuron cell body. A biochemical connection between dendritic spines and the neuron cell nucleus is necessary to maintain a large population of proteins in the postsynaptic compartment. Dendritic spines depend upon the neuron nucleus to enhance their supply of receptors and structural proteins when there is an increase in the number of incoming neurotransmitter signals.

Biochemical molecules synthesized in dendritic spines and later found in the neuron nucleus include certain kinases and transcription factors. Kinase enzymes attach phosphate groups to other proteins. Protein transcription factors, which bind to nuclear DNA, initiate copying of the DNA genetic code into messenger RNA (mRNA). Cells use mRNA as a template for construction of new proteins.

Specific transcription factors initiate reading of the code of individual genes of the DNA. Kinases from the dendritic spines convert transcription factors already in the neuron nucleus from an inactive to an active form. At least 300 genes are thought to be regulated by the biochemical molecules originating in dendritic spines.

It appears the mRNA created in the nucleus in response to synaptic activity at the dendritic spines, and the molecular equipment needed for quick synthesis of protein from that mRNA, transport to dendritic spines in granular vesicles. The location in the neuron where packaging of the mRNA occurs is not known, nor is the nature of the vehicle used to deliver the granules to the dendritic spines identified. However, granular vesicles containing mRNA do travel through neuron cytoplasm to dendritic spines. These granular vesicles remain in a latent protected state until needed.

With increased neurotransmitter activity, mRNA released from the granules commences new protein synthesis causing the dendritic spines to enlarge. When input of neurotransmitter to the postsynaptic compartment slows, free Ca^{++} in the dendritic spine decreases and the process of supplying new mRNA/ribosome granules from the neuron nucleus also slows and the dendritic spine shrinks.

EXCITATORY AND INHIBITORY NEURONS

EXCITATORY GLUTAMATE SYNAPSES

Neurons releasing the neurotransmitter glutamate are *excitatory neurons*. Glutamate synapses display a greater complexity than the generic synapse presented earlier (*Figure 5-2*). This is not surprising because excitatory glutamate neurons support multifaceted aspects of memory formation and memory recall. Glutamate neuron synapses are lost early as dementia progresses.

Glutamate synapses include, in addition to two neurons, cells that are not neurons. Other brain cells forming part of glutamate synapses are glial astrocytes and surveying microglia. Glial astrocytes surround neuron cell bodies and participate in the timely uptake of neurotransmitter from synaptic clefts. Surveying microglia monitors the well-being of synapse activity many times per hour. Chapter 6, *"Introduction to the Glia and Microglia—Meet the Stage Crew"* reviews the role of the glia and microglia at synapses.

There are multiple forms of the glutamate receptor in the brain. Subtypes of glutamate receptors that open ion channels may be separated from each other by their ability to bind pharmaceutical drugs. An added feature at glutamate synapses is the presence of another variety of glutamate receptor that does not open ion channels. Instead these receptors respond to glutamate by activating internal neuron processes. Glutamate receptors forming ion channels localize to the postsynaptic membrane in, and near, the postsynaptic density. The glutamate receptors without an ion channel are found both in pre- and postsynaptic membranes.

Glutamate receptors, which do not form ion channels, bind glutamate on a portion of the receptor extending outside the neuron that causes a change in the position of a part of the receptor located inside the neuron. Repositioning of the cytoplasmic part of this glutamate receptor allows it to bind to cytoplasmic regulatory molecules. The bound regulatory molecules set in motion synthesis of new molecules in the cytoplasm that, depending upon neuron circumstances, may enhance or reduce the effectiveness of the synapse.

Glutamate's ion channel receptors are named AMPA and NMDA after the drugs specified as alpha amino-3-hydroxyl-5-methyl-4-isoxazole-proprionate and N-methyl-D-aspartate, respectively. AMPA and NMDA glutamate receptors work as a team responsible for glutamate's immediate excitatory action at the postsynaptic density. The AMPA receptor responds first.

Mobile AMPA receptor proteins rapidly cycle into and out of the postsynaptic density from adjacent areas of membrane. AMPA receptors include four different protein subunits named GluA1, GluA2, GluA3 and GluA4. Four of these subunits in various combination, most often two GluA1 with two GluA2, cluster together in the postsynaptic membrane creating an ion channel at the center of their complex (*Figure 5-5*).

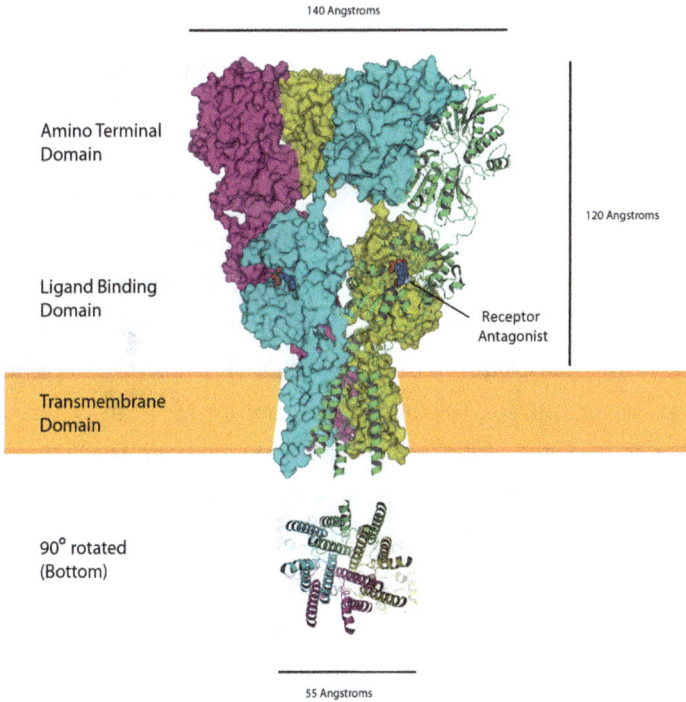

Figure 5-5: The AMPA receptor for glutamate bound to a glutamate agonist showing the amino terminal, ligand binding domain and transmembrane domain. This reconstruction is based upon file 3KG2 in the protein data bank. ©Curtis Neveu

AMPA and NMDA receptors both form non-selective cation channels. *Cations* are ions carrying a positive charge. Non-selective cation channel filters allow simultaneous passage of Na^+, Ca^{++} and some K^+. Na^+ and Ca^{++} enter dendritic spines through these channels and K^+ flows out of the dendritic spine as each ion moves down its concentration gradient. The net result of the larger influx of Na^+ compared to efflux of K^+ is depolarization of the transmembrane potential.

The AMPA and NMDA forms of the glutamate ion channel receptors exist as a pair on glutamate responsive dendritic spines. The AMPA receptor is the first to open in response to glutamate. The NMDA receptor's ion filter is blocked by magnesium ion (Mg^{++}) when the membrane potential is at its normal negative resting value.

NMDA receptor relies upon the AMPA receptor to initiate depolarization of the membrane in response to glutamate. Glutamate binding to the AMPA receptor rapidly causes membrane depolarization forcing Mg^{++} away from the NMDA filter. Glutamate is then able to open the NMDA filter and expand cation flow into the dendritic spine.

The NMDA receptor like the AMPA receptor is a four subunit protein complex composed of two GluN1 and two GluN2 subunits. Different genes code four isoforms of the GluN2 subunit. This diversity allows the human brain to display many variations of glutamate ion channel receptors.

In addition to the prerequisite of membrane depolarization by the AMPA receptor before the NMDA ion channel can open, NMDA receptor requires binding of a second molecule. The second molecule may be either of the amino acids glycine or serine (*Figure 5-6*). Both of these amino acids are probably contributed by the astrocyte glial cells that surround brain synapses. There are also other areas on the NMDA receptor where drug molecules can bind, change its shape and modify its ability to function called *allosteric sites.*

Activated NMDAR

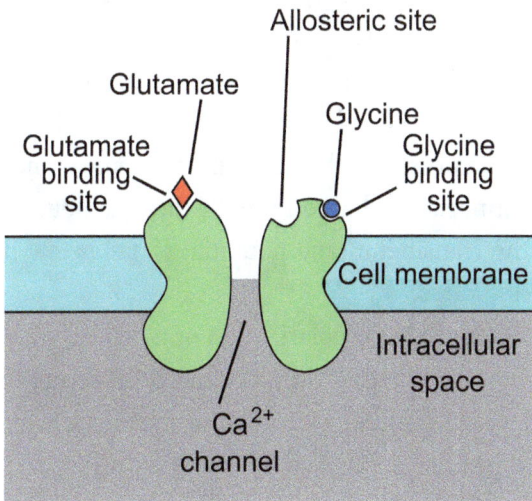

Figure 5-6: Illustration of two of the four subunits of a whole NMDA receptor showing the relative location of the glutamate and the glycine binding sites. Glutamate and glycine do not occupy the same subunit. Four subunit NMDA receptors require the binding of two molecules of glutamate and two of glycine or serine for activation. © CurtisNeveu

INHIBITORY GABA SYNAPSES

GABA is synthesized by enzymes in the brain from glutamate. Neurons using GABA as a neurotransmitter form synapses on the neck of the dendritic spines (*Figure 5-4*) or on the dendrite itself between dendritic spines. GABA neurons also form synapses on the membrane of neuron presynaptic axon terminals, on neuron cell bodies and sometimes at sections of axons without myelin cover. Reuptake of GABA from a synaptic cleft is accomplished by the presynaptic axon terminal with some assistance from astrocyte glial cells in the area.

There are two types of GABA receptors designated GABA type A and GABA type B. GABA binding to a GABA type A receptor opens an ion channel that is selective for chloride ion, Cl^-. Because Cl^- is higher in the interstitial fluid than in neuron cytoplasm, Cl^- diffuses through GABA ion channel receptors into the postsynaptic compartment. The additional negative charge brought into the neuron by Cl^- causes the transmembrane potential to fall below its normal resting value.

Negative hyperpolarization of a neuron transmembrane potential makes it difficult for Na^+ entering through glutamate ion channel receptors to make the membrane potential positive enough to initiate dendritic spikes. Neurons using GABA as a neurotransmitter acquire their generic title, *inhibitory neurons*, because they block the ability of glutamate to perform its normal excitatory purpose, initiation of dendritic spikes.

In contrast to GABA type A receptors, GABA type B receptors traverse the membrane without forming an open channel. When GABA binds to the extracellular portion of a GABA type B receptor, the portion of the protein inside the neuron repositions itself to better facilitate a series of cytoplasmic events. Those cytoplasmic events over time cause some of the neuron's K^+ channels to open and to remain open for a period of time.

The exit of K^+ from the neuron hyperpolarizes the transmembrane potential even further than the entrance of Cl^- alone. The combined influence of the two types of GABA receptors at dendrites eliminates the effect of open glutamate ion channels leading to an absence of dendritic spikes and lack of new structural proteins and receptors for the postsynaptic compartment.

The GABA type A receptor, schematically diagramed in *Figure 5-7*, consists of five subunits arranged around a central channel for Cl^-. Each subunit contains four transmembrane domains. Opening of the Cl^- channel requires binding of two GABA molecules. The size of the filter in the channel can be modified by the presence of drug molecules binding at various allosteric sites.

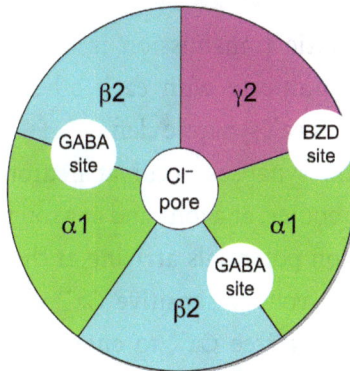

Figure 5-7: Schematic drawing of the GABA type A receptor protein illustrating the five combined subunits forming the Cl^- channel. The positions of the two GABA binding sites and the drug binding site for benzodiazepine (BZD) are illustrated. Illustration is in the public domain

Many pharmaceutical drugs bind to GABA type A receptors and influence their operation. Drug molecules bind at allosteric sites separate from GABA's binding site. The physiologic effects of the drugs include sedation, amnesia and anticonvulsant activity.

OTHER NEUROTRANSMITTERS

All brain synapses present a variation of the basic theme introduced with discussion of glutamate and GABA synapses. Too many neurotransmitter receptor types and synapse configurations exist in the human brain to discuss the details of each of them here. However, like glutamate and GABA, the other neurotransmitters are ligands for receptors that set in motion complex biochemical pathways within neurons.

SUMMARY CHAPTER 5

- Synaptic transmission is used by neurons to communicate with each other across narrow spaces call synaptic clefts
- Brain synapses occur at neuron dendrites, at cell bodies, at axons and at axon terminals
- Action potentials arriving at the axon terminal cause voltage-sensitive Ca^{++} channels to open allowing free Ca^{++} to enter the axon terminal cytoplasm
- Cytoplasmic free Ca^{++} in the axon terminal sets in motion processes leading to the release of a small chemical called a neurotransmitter

- Neurotransmitter activates receptors of the postsynaptic compartment that reside in the postsynaptic density
- Neurotransmitter is removed from the synaptic cleft by reuptake pumps of the presynaptic compartment and by astrocyte glial cells
- Synapses at neuron dendrites occur either on the main shaft of the dendrite or on projections of the dendrite's membrane called dendritic spines
- Neurons using glutamate or GABA as a neurotransmitter control the largest portion of brain activity when a person is awake
- There are two classes of glutamate and GABA receptors in the brain, one class quickly opens ion channels in the postsynaptic density, the other class slowly changes cellular processes within neurons
- Glutamate receptors of the postsynaptic density depolarize the transmembrane potential in the presence of glutamate
- GABA receptors on the dendrite hyperpolarize the transmembrane potential and block the depolarizing effects of glutamate
- Na^+ currents entering the dendrite through glutamate receptors trigger an action potential-like response at the dendrite's membrane called a dendritic spike

- Ca^{++} currents entering dendritic spines set in motion synthesis of additional proteins needed for further development of the synaptic connection
- GABA receptors hyperpolarize the dendrite by opening channels for Cl^- and K^+
- Excitatory neurons release the neurotransmitter glutamate
- Inhibitory neurons release the neurotransmitter GABA

[6]

Introduction to the Glia and Microglia—Meet the Stage Crew

ONLY 10% OF BRAIN CELLS are neurons. The remaining 90% of cells participate as partners of the neurons in management of brain function. The general name for the non-neuron cells in brain tissue is glia. The label glia originates from a word in the Greek language that means glue. The first brain anatomists thought glia supplied scaffolding for neurons. The fragile appearance of neurons suggested a need for some type of physical support. Hence the misnomer implying glia is glue to cement neurons together.

There is also a group of small cells in the brain once thought to be part of the glia. Because of their size they were named microglia. However, observations in recent years established brain microglia is not part of the glia population. The studies showed microglia originates from a different stem cell population than the glia.

STEM CELLS

RADIAL CELLS

In embryos radial cells produce embryonic neurons. Radial cells (*Figure 6-1*), named for their radiating appearance, originate from the cells forming the neural plate during the initial phases of embryonic development. Radial cells can also mature into glia. Mature glia includes two distinct cell populations, astrocytes and oligodendrocytes.

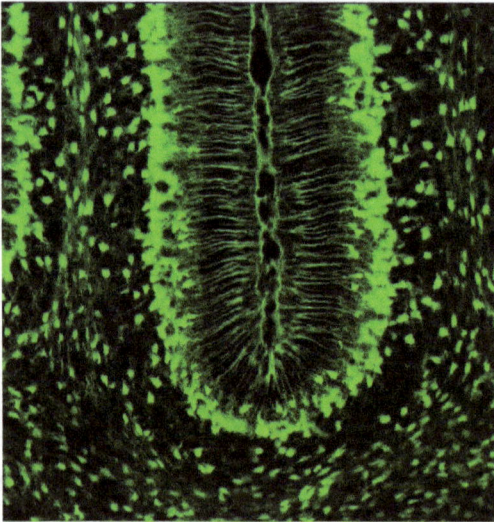

Figure 6-1: Radial glia, also known as Bergmann glia, of the cerebellum of a mouse aged 7 days. This work is in the public domain in the United States because it was produced by an officer or employee of the United States Government as part of the person's official duties.

As radial stem cells begin to mature, a portion of their population stops and does not continue on to full maturity. These partially differentiated cells become the brain's *neural stem cells*. The neural stem cell population persists in juvenile and adult brain and provides a source of replacement neurons and oligodendrocytes.

Neural stem cells capable of becoming new neurons exist in the adult brain under the layer of ependymal cells lining the lateral ventricles and in a zone of the hippocampus region of the cerebral cortex. The hippocampus, discussed in the next chapter, is recognized for its prominent role in memory formation and memory recall.

ASTROCYTES

In many regions of the brain and spinal cord, astrocyte glia develops from radial cells after the original neurons mature. Few astrocytes display measurable rates of cell division in human brain tissue. Replacement of astrocytes seldom occurs. It appears the majority of the original astrocytes survive most of a lifetime.

OLIGODENDROCYTES

In the embryo, radial cells destined to become oligodendrocytes, sometimes also referred to as oligodendrocyte progenitor cells (OPCs) first appear in the ventral forebrain. OPCs travel long distances from the forebrain to reach their ultimate destinations in the developing brain.

Many monitoring molecules support the progress of oligodendrocytes migration through brain tissue to the neurons they will protect. Regulator molecules guide contact-based migration of OPCs over extracellular matrices, axon tracts and astrocyte surfaces. Expansion and retraction of slender membrane extensions of maturing oligodendrocytes aid their migration and enhance their ability to make contact with neuron axons.

A portion of the OPCs distributing through brain does not immediately mature into oligodendrocytes. The progenitor cells remain dormant in the tissue as part of the neural stem cell population until encouraged by chemical signals in the brain to resume development.

Microglia

Microglia originates from the extra-embryonic yolk sac during initial blood cell formation in early embryogenesis. Microglia expresses some but not all of the same antigens as macrophages formed during blood formation by bone. However, microglia seeds the central nervous system before blood formation begins in bone.

Microglia is present at all stages of brain development. In healthy brain, maintenance and local expansion of the microglial population is dependent upon self-renewal of resident cells. All precursor cells for microglial expansion in adult brain arrived there prior to birth.

The discovery that microglia, mature brain cells, retains the ability to self-renew indefinitely in the same fashion as adult stem cells has caused scientists to review their description of stem cells. By definition stems cells are immature cells that can proceed in two directions. They can mature into specialized cells, or they can divide to produce more stem cells.

Dogma until recently stated that mature cells cannot self-renew, that is reproduce themselves indefinitely like stem cells. This property of brain macrophage-like cells to self-renew is also observed in the macrophage population of certain other tissues.

ADULT GLIA AND MICROGLIA

ASTROCYTES

Astrocyte glia comprises the most abundant cell population in the adult brain and spinal cord. Astrocytes outnumber neurons by over fivefold. No region of the brain lacks astrocytes (*Figure 6-2*).

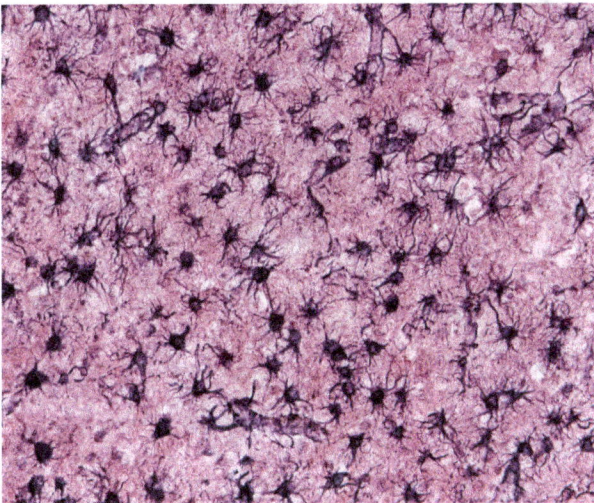

Figure 6-2: Protoplasmic astrocytes in the human cerebral cortex stained with Cajal's gold sublime method. Photomicrograph: ©Jose Luis Calvo

Astrocytes tile the brain in an ordered non-overlapping arrangement. Individual astrocytes manage distinct domains. Only the most distal tips of membrane extensions radiating from individual astrocyte cell bodies touch one another. Each astrocyte exhibits 5-10 main radiating extensions that give rise to numerous finer branches.

Astrocytes divide into two main subtypes, *protoplasmic* and *fibrous*, based upon their morphology and anatomic location in the brain. Protoplasmic astrocytes spread in the gray matter near neuron bodies. Fibrous astrocytes (*Figure 6-3*) position themselves throughout all white matter where they make contact with axons at Nodes of Ranvier.

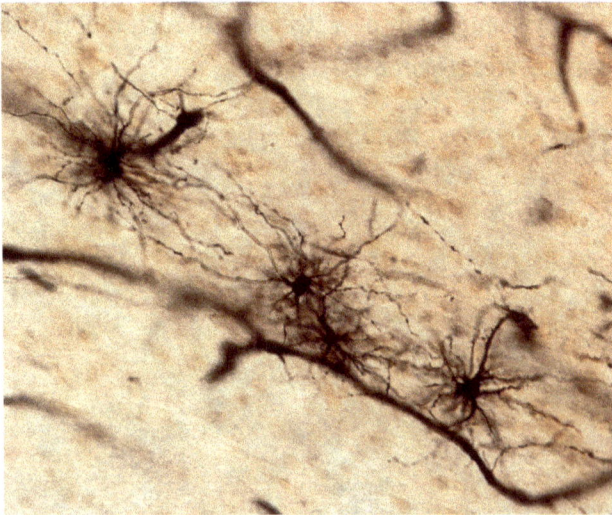

Figure 6-3: Fibrous astrocytes of brain tissue showing many long thin membrane extensions stained by the Golgi silver chromate method. Photomicrograph: ©Jose Luis Calvo

Both types make extensive contact with blood vessels and regulate blood flow by secreting molecules to either dilate or constrict the brain capillaries (*Figure 6-4*).

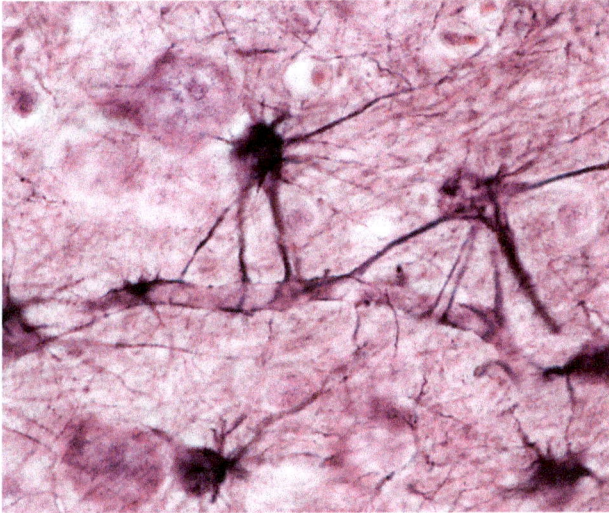

Figure 6-4: Astrocyte membrane extensions with end feet to the walls of a blood vessel stained with Cajal's gold sublime method. Photomicrograph: ©Jose Luis Calvo

OLIGODENDROCYTES

Oligodendrocytes located at the axons of neurons produce the white color marking neuron tracts in the brain. White matter of the adult brain is crowded with both mature oligodendrocytes and oligodendrocyte progenitor cells. Oligodendrocytes experience a brief period of time for myelination as they mature from oligodendrocyte progenitor cells to oligodendrocytes. At full maturity, oligodendrocytes become incapable of myelination of additional axons.

Formation of a wrap on multiple axons by solitary oligodendrocytes is a coordinated event. They select axons with diameters greater than 0.2 micrometer (2×10^{-7} meter) and often envelop about 60 axons simultaneously. Upon contact with neuron axons, the exploratory membrane extensions convert to flat sheets that spread and wind around neuron axons to create a multilayered or laminar cover. The laminar structure then compacts extruding almost all of its cytoplasm and becomes myelin.

Due to the complexity of human brain organization, myelination of the frontal cortex takes decades. Myelination of neurons in the human brain starts in late fetal life and the rate of myelin formation peaks in infants. Yet, certain regions of the cerebral cortex, neurons that control ongoing activities of life, continue to increase their myelin into the 5th and 6th decade of life.

MICROGLIA

Microglia represents 5-20% of the total non-neuron population and is distributed in all brain regions. Numbering approximately one per neuron in mammalian brain, microglia arranges itself in a non-overlapping grid. Each surveying microglia monitors the area around a neuron. Microglia is present in both white and gray matter.

In healthy states, *surveying microglia* exhibits a branched morphology. Its slim membrane extensions intermingle with neurons and astrocytes. The cell body of surveying microglia is not mobile (*Figure 6-5*). Only the long extensions move and assess the quality of the local environment.

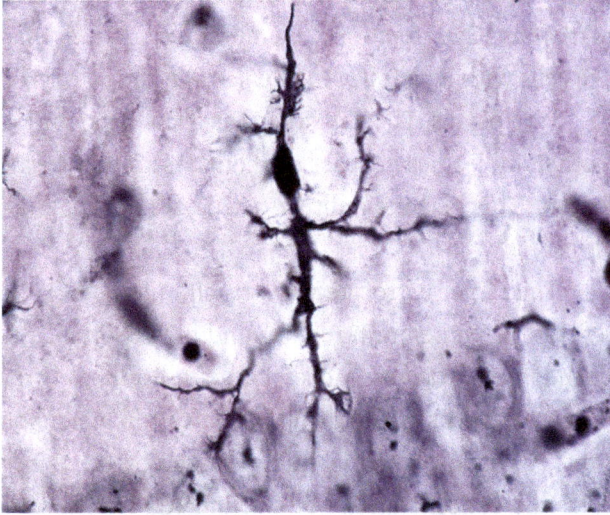

Figure 6-5: Surveying microglia stained with Rio Hortega's silver carbonate method. Photomicrograph: ©Jose Luis Calvo

FOUR PART SYNAPSES

The previous chapter reviewed the two parts of a neuron synapse, the presynaptic compartment and the postsynaptic density. In real brains, however, the synapse where neurons come together is a four part structure. The four-part synapse includes the presynaptic and postsynaptic neurons, an astrocyte network and surveying microglia. The crowded collection of several cell types around neuron synapses presents a diversity of communication opportunities for maintenance of healthy brain function and for networking among cells.

ASTROCYTE WEBS

Astrocytes, connected by *gap junctions* where the tips of their membrane extensions come in contact, create cellular webs. Gap junctions, membrane structures shared by two cells with *a central open channel,* join the cytoplasm of multiple astrocytes. The gap junctions allow ions, nutrients and other molecules to pass without restriction between astrocytes. This permits multiple astrocytes to behave like a single cell.

The larger astrocyte webs reach millimeter (1×10^{-3} meter) size in some brain regions and include hundreds to thousands of astrocytes. Movement of substances through astrocyte networks is modified by activity in neighboring neurons. Neuron signaling initiates changes in astrocyte cytoplasmic concentrations of calcium (Ca^{++}) and sodium (Na^+), powerful ions for triggering cytoplasmic responses.

Astrocyte networks organize themselves into separate anatomic compartments. Astrocyte networks near the neurons committed to a unique functional activity may experience an influential relationship exclusive to that set of neurons. Alternately, astrocyte networks may act as barriers to isolate neuron systems and restrict information transfer between neuron circuits. Neuron circuits include multiple neurons performing as a single unit. Astrocyte barriers may provide effective segregation of neuron activity in various parts of the brain.

Boundaries formed by both protoplasmic and fibrous astrocytes take part in guiding the migration of oligodendrocyte progenitor cells. And, both subtypes of astrocytes provide physical and functional support at brain capillaries. Astrocytes maintain integrity of the blood brain barrier and regulate blood flow to match it to neuron requirement for glucose and oxygen. Protoplasmic astrocytes link neuron bodies and dendrites with blood vessels. Fibrous astrocytes link axons with blood vessels.

SURVEYING MICROGLIA

Surveying microglia, sometimes called resting microglia, possesses a tiny stationary cell body (*Figure 6-5*). The membrane extensions emanating from the body elongate and retract as they explore the brain tissue environment. Surveying microglia scans its local area with precision and makes physical contact with synapses to monitor synaptic activity.

Surveying microglia makes brief, about 5 min, contacts with neuron synapses at a frequency of about once per hour. The timing of the contacts depends on the frequency of neurotransmitter release by the presynaptic compartment. Surveying microglia contacts synapses less often when neuron activity lessens.

Surveying microglia makes contact with neurons at synapses on dendritic spines but not with the shafts of dendrites. Its contact at synapses may be mediated by a variety of chemo-attractant and repellant signaling molecules. Proposed signaling molecules include glutamate, ATP, molecules named chemokines, brain-derived growth factor and ion currents.

FUNCTIONAL PARTNERSHIPS

ASTROCYTES AND NEURONS

The extent of astrocyte engulfment of neuron synapses and Nodes of Ranvier varies by brain region. However, all synapses in the brain maintain some interaction with astrocytes.

In areas of the brain like the hippocampus and cerebral cortex, the many branches of a single protoplasmic astrocyte may contact several hundred dendrites from multiple neurons and envelop up to 100,000 or more synapses. This affords individual astrocytes as well as astrocyte networks extensive neuron monitoring potential.

A primary function of fibrous astrocytes is to monitor and maintain steady ion concentrations in the interstitial fluid around axons. Their extensive contact with blood capillaries permits them to correct any deviation from the acceptable range of interstitial fluid ion concentration, which is essential to action potential propagation.

Astrocytes, like postsynaptic compartments of neurons express neurotransmitter receptors on their distal branches. The neurotransmitter receptors expressed match the neurotransmitter receptors of the neuron synapses they contact.

Yet, the timing of ion currents at the astrocyte bodies indicates neurotransmitter receptors on their branches do not respond to normal levels of neurotransmitter. Under stressful conditions, astrocytes possess the potential to release the neurotransmitter to activate their own receptors. In healthy brain, however, this is not thought to occur.

Neurons, which are active when the brain is awake, release a neurotransmitter named glutamate. Glutamate is removed from synaptic clefts by resident astrocytes. The cell membrane of astrocytes near a glutamate synapse expresses transporter proteins for glutamate.

The drain of glutamate from neurons is compensated for by flow of a molecule named glutamine from astrocytes to neurons. About one third of the glutamate taken up by astrocytes is converted to glutamine and returned to the neuron. Within the neuron glutamine is converted back to glutamate and repackaged into vesicles for synaptic release.

Astrocytes supplement glutamine made from neuron glutamate with glutamine synthesized *de novo* from glucose. The remaining two thirds of the glutamate taken up by astrocytes replenish components of the astrocytes mitochondrial ATP-producing cycle. Neurons lack the enzymes necessary to make glutamate from glucose.

One theory is this arrangement provides an energy advantage to neurons. It permits ALL the glucose entering neurons to be moved with high efficiency through glycolysis and ALL the resulting pyruvate to be transferred into their mitochondria to produce high energy molecules of ATP. If neurons made glutamate, part of their pyruvate from glucose metabolism would be drawn off for glutamate synthesis and would result in a lower overall ATP production.

OLIGODENDROCYTES AND NEURON AXONS

The thickness of an axon's myelin sheath varies. Variation in myelin thickness along with differences in the length of individual Nodes of Ranvier may provide additional opportunities for neurons to adjust the speed of their action potentials. Because only larger axons attract oligodendrocytes, it is assumed electrical activity of the axon is an essential signal for myelination. In support of this idea, when intricate motor skills are learned, axon myelination increases in synchrony with increased motor cortex neuron activity.

Neuron axon transport systems and overall neuron viability also appear to depend upon supportive signals from oligodendrocytes. Oligodendrocytes insure the well-being of axons by releasing growth stimulating molecules including glial cell neurotropic factor, brain-derived neurotropic factor and insulin-like growth factor. Axons with damaged myelin sheaths display alteration in the stability of their cytoskeleton which leads to adverse changes in the transport rates of material between the neuron body and the axon terminal.

Oligodendrocytes may play a larger role in establishing electrical circuits in the brain than is appreciated. Proteins in the oligodendrocytes' membrane inhibit axon collateral formation. Observed pattern variation in the length of the Nodes of Ranvier, therefore, may affect a neuron's opportunities for using axon collaterals in its circuitry design. In addition, synapses on axons occur only where an axon is bare at the Nodes of Ranvier.

Microglia and Neurons

Microglia connects with astrocytes and neurons at synapses. There is a ligand for a brain receptor on the membrane of neurons, astrocytes and oligodendrocytes named CD200. The receptor for CD200 is exclusive to microglia. The CD200 molecules on the surface of neighboring cells keep microglia in its surveying state. Surveying microglia also expresses receptors for most brain neurotransmitters that allow it to monitor activity at all brain synapses.

The behavior of surveying microglia is regulated and guided by activity of individual neurons. Neurons with high levels of activity release ATP. ATP in the synapse induces microglia to polarize its extensions toward these dynamic synapses. Surveying microglia removes dendritic spines when neuron activity slows. Pruning of synapses receiving little neurotransmitter input is required to maintain the dynamic character of neuron circuits. In the developing visual system, live images observed with confocal microscopy followed by electron microscopy confirm surveying microglia removes synaptic elements.

Metabolism in the Brain

Glucose for Energy

The brain relies on blood glucose as its primary source of energy. Glucose moves into the brain through capillary endothelial cells with the help of transporter proteins. Astrocytes whose extensions reinforce the blood brain barrier are well positioned to take up entering glucose. Neurons also absorb incoming glucose. To assist neurons in this endeavor, astrocytes secrete molecules that dilate blood vessels. Astrocytes adjustment of blood flow matches activity at synapses to availability of glucose (*Figure 6-4*).

Glucose can be stockpiled in some cells in the form of a storage molecule named *glycogen*. Astrocytes serve as the principle storage site for glycogen in the brain. Glycogen containing astrocytes exist in areas where synapses pack close together. Astrocyte glycogen stocks support normal neuron activity in times of low blood glucose. Exceptionally active neurons create areas with low blood glucose. When blood glucose becomes inadequate, glycogen is converted first to glucose and then metabolized to lactate by astrocytes for delivery to neurons. The neurons convert lactate to pyruvate to fuel their production of ATP.

When brain activity is in its normal range, the astrocyte stockpile of glycogen is gradually broken down and restored. The routine uptake of glutamate neurotransmitter by astrocytes triggers conversion of glycogen to glucose. In addition to using glucose to make glutamine to send to the neurons, astrocytes make ATP to support their own Na^+ ion pumps. When astrocytes remove glutamate from a synaptic cleft using transporters, Na^+ is brought into the cell along with glutamate. Ion pumps must then restore astrocyte cytoplasm to its normal low level of Na^+.

Brain cells, other than neurons and astrocytes, also use glucose for energy. Oligodendrocytes during the myelination procedure consume substantial amounts of glucose, oxygen and ATP. Surveying microglia is not thought to require a high rate of glucose utilization, but *activated microglia*, needed to repair brain damage or stave off infection, increases its glucose uptake.

CHOLESTEROL FOR MEMBRANE STRUCTURE

Many different kinds of lipids, fat molecules, are used to form brain cell membranes. Some lipids cross into the brain from blood, others cannot navigate through the blood brain barrier. Cholesterol is among the lipids that cannot cross the blood brain barrier. There is no mixing of brain cholesterol with cholesterol in blood. All cholesterol used by brain cells is made in the brain.

The brain contains more cholesterol than any other organ. Twenty to twenty-five percent of a person's cholesterol is synthesized in the brain and resides there. Cholesterol is a major component of brain cell membranes (*Figure 6-6*).

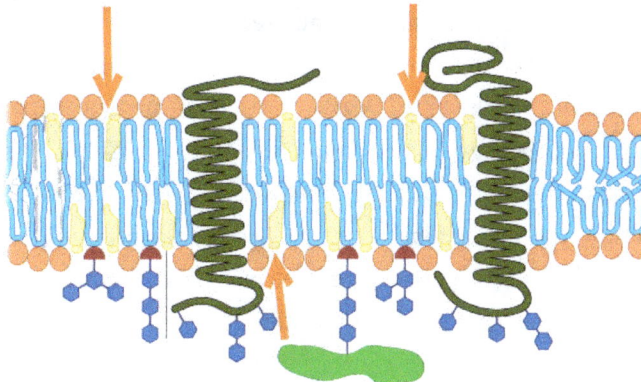

Figure 6-6: Illustration of cholesterol's role in lipid membrane formation. Arrows point to cholesterol (yellow) inserted between structural lipids in a cell membrane. Illustration: ©WarX, released to the public domain

Cholesterol is critical to synapse formation, dendrite formation, axon elongation and is a major component of myelin. Brain cholesterol is also vital to efficient neurotransmitter release from axon terminals. It is an important component of vesicles used for transport of materials within neurons along cytoskeletal components. Cholesterol makes regions of membranes less fluid and more resistant to the passage of charged molecules. It presence at synapses supports the function of ion channels and membrane receptors.

Astrocytes and oligodendrocytes produce 2-3 times more cholesterol than neurons. Astrocytes provide cholesterol for growth of neurons and to support synapse structures. The concentration of cholesterol in brain is kept stable. Its half-life is reported to be as long as 5 years. In contrast, in blood plasma the half-life of cholesterol is a few hours.

Brain cells take up cholesterol enclosed in *apolipoprotein E (ApoE)* with a membrane receptor. ApoE, produced by astrocytes and microglia, is the major transporter of cholesterol within the brain. Astrocytes secrete small ApoE particles composed of phospholipid and cholesterol that perform like high-density lipoprotein (HDL) circulating outside the brain. The larger triglyceride-containing low-density lipoprotein (LDL) particles produced by liver appear absent in the central nervous system.

Elimination of cholesterol from the brain requires its conversion to 24S-hydroxycholesterol. This enzymatic change occurs mainly in neurons. The conversion product crosses the blood brain barrier and exits neurons and the brain without restriction.

REPAIR OF BRAIN DAMAGE

REACTIVE ASTROCYTES

Neurons respond to areas of injury in the brain by adjusting their circuits in undamaged sections of the brain to by-pass failure of the damaged tissue. Isolation of injured tissue followed by tissue repair or elimination of damaged tissue is left to astrocytes, oligodendrocytes and microglia.

Astrocytes respond to all types of central nervous system injury. Astrocytes reacting to tissue rupture remove excess glutamate. This is protective because surplus glutamate in brain fluids is toxic to neurons. Astrocytes shield cells in damaged brain tissue from oxidative stressors by producing the anti-oxidant glutathione. Glutathione neutralizes highly destructive oxygen-containing molecules released by the injury. Astrocytes also facilitate repair of damage to the blood brain barrier and reduce excess fluid accumulation, edema.

A response known as *reactive astrogliosis* is a hallmark of structural damage in the brain. Reactive astrogliosis is a graded response. Small injuries produce a mild to moderate form of astrogliosis. Larger injuries produce severe diffuse astrogliosis with or without glial scar formation.

During mild to moderate astrogliosis some astrocytes display enlargement of their body and extensions. There is no extensive reorganization of the tissue structure. This form of astrogliosis is associated with mild, nonconcussion trauma and viral infections. With resolution of the injury, astrocytes return to their normal state.

In contrast, severe diffuse astrogliosis is found surrounding large focal lesions and infections and in those brain areas experiencing neuron deterioration. The volume of astrocyte bodies and membrane extensions enlarges. Astrocyte extensions spread beyond their normal domain and overlap those of their neighbors. This type of astrogliosis produces lasting reorganization of the tissue, but it does not produce scar tissue (*Figure 6-7*).

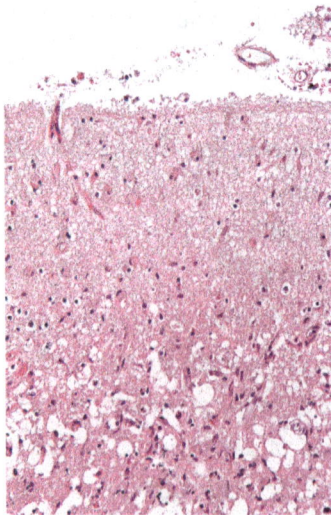

Figure 6-7: High magnification photomicrograph showing reactive astrocytes restructuring tissue in this image of ischemic brain. Photomicrograph: ©Nephron

Astrocyte response to highly destructive insults to the brain including tissue penetration by a foreign object, blockage of blood flow and torn tissue includes all of the features of severe diffuse astrogliosis plus formation of dense, compact glial scars. Scars form along the edges of tissue destruction. Astrocytes interact with oligodendrocytes, microglia and cells of the meninges to deposit dense collagen. The deposited material contains molecular signals to inhibit neuron regeneration within the damaged area. In this form of astrogliosis, structural changes in the brain tissue last long after the injured section is enclosed.

AMOEBOID MICROGLIA

Surveying microglia responds to mechanical brain injury by changing to its macrophage-like form. Torn brain tissue releases ATP from neurons and other damaged cells. Surveying microglia travels toward released ATP and then makes an executive decision whether neurons may be salvaged or not. If the neurons cannot be repaired, microglia assumes its macrophage-like form, releases lethal compounds to finish the kill and removes debris by engulfing it.

For salvageable neurons, microglia releases factors to assist in their repair and recruits oligodendrocyte progenitor cells. The restoration of myelin to damaged axons cannot be performed by existing mature oligodendrocytes. Instead oligodendrocyte progenitor cells, distributed throughout the brain, must be persuaded to mature. The transition of oligodendrocyte progenitor cells into the maturation sequence appears to be initiated by factors secreted by activated astrocytes and microglia that draw them to the damaged site.

INFLAMMATION AND INFECTION

Surveying microglia is the first responder to infection in the brain. When microglia detects pathogens with its membrane receptors, which resemble the receptors on immune cells outside the brain, it becomes amoeboid and it multiplies. Then, like macrophage, microglia engulfs and destroys the pathogen. Pathogen-activated microglia releases nitrous oxide, charged oxygen molecules, pro-inflammatory cytokines, protein destroying enzymes and toxic levels of glutamate. Activated microglia also interferes with astrocyte uptake of glutamate, further increasing destructive glutamate in the tissue.

Inflammation in response to viral and bacterial infection in the central nervous system is different than classical inflammation elsewhere in the body. It does not result from the entry of immune cells from blood. It does not produce heat or edema in the tissue. Rather, activated microglia and reactive astrocytes produce molecules similar to those released by immune cells outside the brain to destroy the pathogen (*Figure 6-8*).

Astrocytes make and secrete molecules with either pro- or anti- inflammatory potential depending upon the brain's situation. In response to pathogen they secrete pro-inflammatory molecules to assist microglia in destruction of the pathogen. Astrocytes can be infected by viruses and act as reservoir for several viral infections. The precise role of astrocytes in viral infection remains unclear.

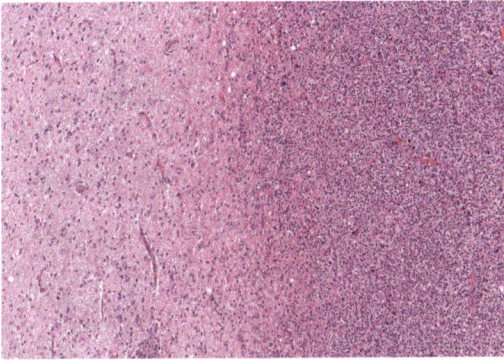

Figure 6-8: Photomicrograph of brain tissue stained with hematoxylin and eosin showing viral encephalitis inflammation on right and normal brain tissue on left. Notice the large increase in dark staining cell nuclei on the right. Photomicrograph: ©vetpathologist

Virus also infects and replicates in microglia. Virus-infected microglia and astrocytes produce molecules toxic to neurons. Death of neurons during viral infection appears to result from an erroneous exchange of signal molecules between astrocytes and microglia in gray and white matter.

The description, inflammation, is expanded in the central nervous system to include diseases not related to infection. In the brain, inflammation produced by degenerative diseases is not part of a healing method but rather a chronic malfunction of a normal protective mechanism. There will be additional discussion of the inflammation of degenerative brain disorders in the final chapter of this book, Chapter 8, "*When It All Goes Wrong—Alzheimer's Dementia.*"

SUMMARY CHAPTER 6

- Radial cells of the embryo mature into neurons, oligodendrocytes and astrocytes
- The portion of radial cells that fails to completely mature during embryonic development persists in adult brain as neural stem cells
- Neural stem cells develop into replacement neurons and new oligodendrocytes
- Astrocytes present at birth survive most of a lifetime
- Neural stem cells capable of maturing into oligodendrocytes exist throughout the adult brain
- Microglia does not originate from radial stem cells but rather migrates to the neural plate from the extra-embryonic yolk sac prior to blood formation in bone
- Astrocytes tile the brain in an ordered arrangement with no region of the brain left uncovered
- Astrocytes, connected at the tips of their long extensions by gap junctions, create large cellular webs
- Due to the complexity of the human brain, oligodendrocytes myelination of neuron axons continues into the 5th and 6th decade of life
- Each surveying microglia monitors the area around one neuron

- Fibrous astrocytes monitor and maintain steady ion concentrations in the interstitial fluid around axons
- Protoplasmic astrocytes remove glutamate from neuron synapses and return glutamine to the neuron for conversion back to glutamate
- Oligodendrocytes myelination of neuron axons insulates, speeds action potentials, provides growth factors and affects synapses on axons and axon collateral formation
- Microglia expresses receptors for most neurotransmitters permitting microglia to monitor neurotransmitter release at all brain synapses
- Brain cells depend on glucose for their primary energy source
- The storage form of glucose, glycogen, is only found in astrocytes
- All cholesterol in brain is made in the brain
- Brain cholesterol is critical to synapse formation, dendrite formation, axon elongation and is a major component of myelin
- The response of astrocytes to brain injury is named reactive astrogliosis
- Activated microglia in its macrophage-like form eliminates damaged neurons
- The first responder to pathogen in the brain is microglia

- Astrocytes secrete pro-inflammatory molecules to assist microglia in destruction of the pathogen
- Inflammation in response to viral and bacterial infection in the central nervous system is produced by pro-inflammatory molecules secreted by astrocytes and microglia

[7]

Brain's Infrastructure for Memory and Language

PSYCHOLOGY STUDIES DISCOVERED many clues about how human memory works over the past 40 years. Yet, mechanisms essential for consciousness and a sense of self remain unclear. When reading this chapter, keep in mind much of the presentation is still theory. One way to think about descriptions of how a brain works is to visualize the particulars as a fuzzy snapshot with too few pixels. Contemporary science recognizes some fundamental truths about how the brain works, but a complete picture remains elusive.

Neuroscientists agree on the general details of how neurons communicate with each other. But, few undisputed facts support the many theories of how neurons create something as intangible as thoughts. Both genetics and a person's environment contribute to self-awareness, however little is known about how the brain's infrastructure supports an individual's perception of the world.

The dogma that the adult human brain cannot replace its neurons was reversed in 1998 by evidence of neurogenesis in the hippocampus of a cancer patient. Incorporation of emerging knowledge of neurogenesis and plasticity of neuron connections into theories of human awareness requires reinterpretation of older data. Also, the part played by the brain's glial cells in neuron signaling is beginning to be appreciated. Human memory may not be the exclusive domain of neurons working in isolation.

Modern speculation occasionally proposes brains operate like computers. Yet, contemporary studies of the central nervous system imply brains work different than typical computers. Most computers calculate in a linear sequence solving problems one step at a time. Brains solve multiple problems and engage in complex environmental analysis all at the same time. They do this with little error in a watery structure tens of millions of times slower than state-of-the-art digital computers.

In brain models of the last century, neurons were imagined to signal in a sequential 'on' and 'off' manner. However, newer models of neuron signaling, which are based upon emerging evidence from physiology and psychology, include interconnected processing units operating in parallel to generate graded and variable signals. With parallel operating models, a small change in one unit adjusts a lot of information in the system at the same time.

Designers of new computers are working to copy the brain's parallel operating organization into their systems.

This chapter presents some of the major hypotheses built around existing evidence of the brain's mechanism for forming memories. Most of the neuroscience discussed comes from investigations conducted in the last 15 years and was obtained with newer non-invasive measuring devices developed for human medical testing. Language acquisition, one form of memory, is presented as an example of the complexity of human memory systems. Present understanding of language acquisition permits relevant, if incomplete, discussion of the neuron circuits that process sensory information to create a form of memory unique to humans.

This chapter first considers neuron coding of information, anatomic patterns of neuron connections and non-invasive tools used to link activity of human brain circuits with specific tasks. Then data obtained with these tools will be used to outline recent ideas about memory formation and language acquisition.

INFORMATION FLOW

NEURON CODES

The original hypothesis stating that signaling between brain neurons is coded by strings of 'on' and 'off' input emerged during studies of sensory input to the brain. The presence of external stimuli including light, sound, taste, smell and touch cause neurons of sensory organs to fire sequences of action potential in various temporal patterns. Many studies conclude the pattern of action potentials entering the brain from a sensory organ encodes information about the sensory stimulus. An extension of those conclusions assumes neuron signaling within the brain is coded in a similar way.

Analysis of sensory action potential patterns uses statistical methods and probability theory. An intense debate continues about whether sensory neurons use rate coding or temporal coding. Rate coding of action potential patterns assumes information about the sensory stimulus is contained in the firing rate of the neuron. Temporal coding assumes precise timing of each action potential in the train contains the information. Acquiring data that separates these theories is a challenge because individual neurons display background firing activity equivalent to noise.

In contrast, newer theories of information processing within the brain hypothesize action potential coding is carried out by populations of neurons working together. This model claims groups of neurons, neuron circuits, perform better than individual neurons because noise in the group gets averaged out. This is frequently referred to as *ensemble coding*.

Other hypotheses claim balance between the average amounts of depolarizing and hyperpolarizing synaptic currents is a basic form of brain information processing. These hypotheses are consistent with the rhythms in brain electrical activity observed with scalp electrodes. If the balance model is correct, any shift in the equilibrium toward net excitation or net inhibition may allow multiple sensory inputs to be clarified as one all-inclusive message at cerebral cortical neurons. Possibly, an alignment of brain electrical rhythms with ensemble coding contributes to quick distribution of information throughout the brain.

RHYTHMIC BRAIN WAVES

Brain neurons are believed never to operate in isolation. Rather, they perform as members of organized groups labeled *neuron circuits*. Each neuron circuit is thought to process specific kinds of information.

The flow of action potentials, moving from neuron to neuron because of synaptic activity, is often compared to electrical current in household circuits. Electrical engineers describe a circuit as a network consisting of a closed loop that provides a return path for an electrical current, often electrons flowing through a wire.

The following discussion of brain activity will use the term 'circuit' to mean a neuron path within the brain where ion flow creates a pattern of action potentials and dendritic spikes. Unlike household circuits for electricity, brain circuits do not form closed loops where action potentials travel back to their points of initiation.

Neurons come in many sizes and shapes. Neuron circuits incorporate the large neurons described in Chapter 4, *"Neurons—How They Make Electricity"* plus many smaller neurons identified as *interneurons*. Interneuron axons and dendrites often look alike because of their small size. Another name for interneurons is association neurons or relay neurons. Interneurons contribute greatly to the observed electrical oscillations known as *brain waves.*

Interneurons influence the output of neuron circuits because they regulate the activity of the larger neurons. They fine-tune the characteristics of a brain's ongoing electrical activity. Most interneurons release a neurotransmitter that hyperpolarizes the presynaptic and/or postsynaptic transmembrane potential at neuron synapses. Membrane hyperpolarization by inhibitory interneurons slows the flow of action potentials, and therefore current, through a neuron circuit.

The total ongoing neuron activity in the brain sums into rhythmic patterns. Sometimes the pattern is the result of synchrony in action potentials arriving at the synapses of a group of neurons. When action potentials of large groups of neurons synchronize they cause large amplitude fluctuations of the local field potential. The local field potential is created by opening of membrane ion channels at synapses and the resulting net flow of ions into neurons from the interstitial fluid and out of neuron cytoplasm into the interstitial fluid.

Fluctuations of local field potentials also appear because of connections between distant brain areas. Some brain areas form feedback circuits by linking together in a circular pattern. Timing of action potential repetitions in various parts of a feedback loop's circuits causes oscillations in regional field potentials. Electrodes placed on the surface of the head measure shifts in local field potentials.

Voltage measurements, the differences in the field potential between pairs of scalp electrodes, produce an oscillatory recording over time named an electroencephalogram (EEG). An EEG measures the synchronous activity of millions of neurons at a time (*Figure 7-1*).

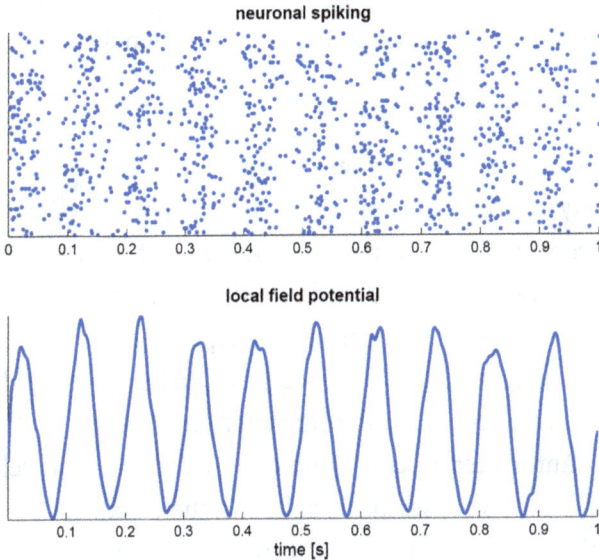

Figure 7-1: Simulation of brain activity showing 10 Hz oscillations. Neuron spiking, action potentials arriving at the presynaptic compartment, is simulated by a rate-modulated Poisson process (upper panel). Local field potential is simulated as the sum ion flow caused by the neuron spiking, representing the mean activity of a large number of neurons (lower panel). Illustration: ©TjeerdB

MAPPING THE BRAIN'S NEURONS

HUMAN CONNECTOME

The term *connectome* is another of science's made-up names. The 'connect' part alludes to mapping all the neuron connections throughout the brain. The '-ome' part is used to suggest a comparison to the genome project that mapped all of an individual's genes. Some like to compare the brain connectome to a wiring diagram. On a micro scale it describes the characteristics of neuron connections surrounding a synapse. On a larger scale the brain connectome is a description of neuron connectivity between distant sections of the brain.

Defining the entire connectome of the human brain is not feasible at present. However, data collected from tiny volumes of brain tissue using automated electron microscopy are confirming the complexity of neuron connections. Brain tissue is sliced into thin serial sections before processing for electron microscopy. The microscopy sections, reconstructed after analysis, produce a three dimensional model of neuron connections in the tissue.

Automated electron microscopy data confirm cells around brain synapses pack tight together. Yet, the presence of synapses cannot be inferred by where axons and dendrites intersect. How individual variation will affect the outcome of ongoing connectome projects remains uncertain. But, automated electron microscopy brain maps, a trillion times finer than those obtainable with external imaging techniques, add essential information.

DIAGRAMS OF THE CEREBRAL CORTEX

The first brain region mapped by anatomists of the nineteenth century is the *cerebral cortex*. Based upon expansion of brain regions in mammalian species through evolution, the most recently developed part of the cerebral cortex is named the *neo-cortex*. The ancient part of the cerebral cortex is the *hippocampus* (*Figure 7-2*).

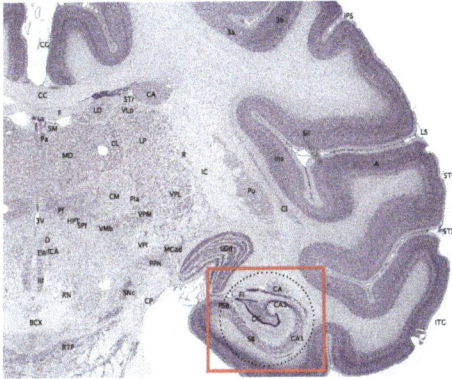

Figure 7-2: Brainmap of a macaque monkey displaying the dark outer band of neo-cortex. The oldest phylogenetic area of the cortex, the hippocampus, is the circled region in the red box at the bottom. Illustration: ©brainmaps.org

In humans as in monkeys, the hippocampus lies under the neo-cortex in the medial temporal lobe. It contains five regions, dentate gyrus, CA1, CA2, CA3 and CA4. The hippocampus codes memory into neuron signaling patterns. It consolidates short-term memory into long-term memory and guides a person's navigation of space.

Anatomists of the late 19th century were intrigued by the layered arrangement of cortex neurons drawn by Santiago Ramón y Cajal and others (*Figure 7-3*).

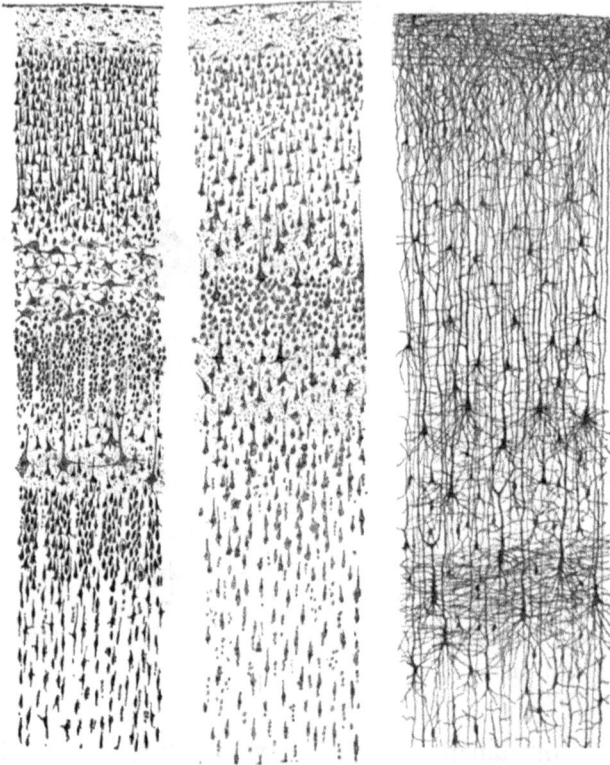

Figure 7-3: The exterior surfaces of the cortical sections are located at the top of the illustration. Left: Nissl stained neuron bodies of the visual cortex of an adult human. Middle: Nissl-stained motor cortex of an adult human. Right: Golgi stained cortex of a 1 ½ month old infant displaying neuron bodies, dendrites and axons of a random 1-3% of neurons present. Notice how the pattern of neuron bodies, dark spots, varies. Drawings: By Santiago Ramón y Cajal found in the book "Comparative Study of the Sensory Areas of the Human Cortex." This work is in the public domain in the United States.

The large neurons of the neo-cortex occupy six layers from exterior surface to interior surface. Neurons from various layers connect to form small microcircuits in cortical columns. The human cerebral cortex is only 2 to 4 millimeters thick. In spite of its size this small band of tissue plays a critical role in language, memory, attention, awareness, thought and consciousness.

Anatomists noticed neuron bodies and neuron axons display diverse patterns of arrangement in various regions of the neo-cortex (*Figure 7-3*). The configurations show marked differences in neuron organization and definite borders where one pattern changes to another design. Each area defined by one of the patterns, when stimulated with electrical current, produces a distinct response.

The German anatomist, Korbinian Brodmann, was the first to map neuron patterns of the entire neo-cortex. He used Nissl stain to locate neuron bodies. Brodmann numbered his areas 1 through 52. Because of his interest in evolution of the neo-cortex, he studied the neo-cortex of humans in postmortem specimens and compared patterns he found with those of eight other species. He discovered some brain areas, Brodmann areas 12, 13, 14-16 and 48-51 in lower mammals, are absent in humans.

Many of Brodmann's areas were renamed when evidence of their purpose emerged. For example, Brodmann area 17 is now recognized as the primary visual cortex and Brodmann area 4 is the primary motor cortex. Modern techniques subdivide many of Brodmann's areas based upon smaller neuron groups with distinctive responses to electrical stimulation. Although refined and renamed for over a century, Brodmann's areas remain the best known and most often cited when discussing the cerebral cortex (*Figure 7-4*).

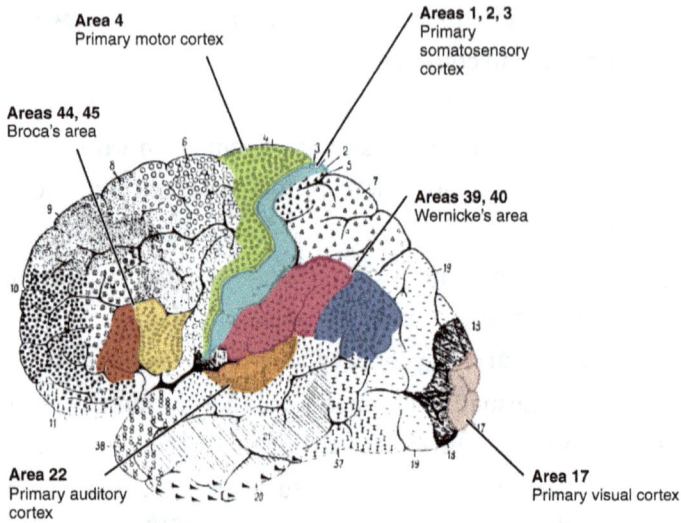

Area 4
Primary motor cortex

Areas 1, 2, 3
Primary somatosensory cortex

Areas 44, 45
Broca's area

Areas 39, 40
Wernicke's area

Area 22
Primary auditory cortex

Area 17
Primary visual cortex

Brodmann's cytotechtonic map (1909):
Lateral surface

Figure 7-4: Numbers refer to areas where Brodmann detected a pattern of neuron bodies with characteristics different than those of the surrounding area. Colored areas correlate Brodmann areas with later data revealing the responsibility of particular neuron groups. Illustration: ©OpenStax College

Under the cerebral cortex deep in white matter, neuron bodies collect into groups. Each group is called a nucleus. In preserved brain, nuclei appear as islands of gray matter within white matter. Neurons included in each brain nucleus form connections with neurons in other nuclei and with neurons in the cerebral cortex.

Larger brain nuclei possess descriptive names because their role in brain activity is extensively described. Modern methods routinely add additional small nuclei to the human brain map. The purpose of most of the multitude of small nuclei once too little to be noticed is still being investigated.

NEURON AXON TRACTS

A traditional method for mapping neuron connections between brain nuclei and cortical areas relies upon injection of a labeling agent, often a dye. This technique is described as *tract tracing*. Tract tracing is considered the gold standard for charting the path of neurons through brain tissue.

Following uptake of dye, serial sections of tissue are examined using electron microscopy to determine location of the dye within neurons. The electron microscopy approach works well over short distances with large neurons but is less than optimal for neuron projections spanning longer distances.

Anatomists investigated brain neuron pathways using tract tracing for over 40 years. Tract tracing requires the axon's cytoskeletal transport system to move the dye from where it enters the neuron to the place where it ends up over a period of time.

Anterograde tracing involves introduction of tracer molecules into neuron bodies and detecting the tracer in the axon terminals. *Retrograde tracing* is achieved by using a labeling agent that is taken up by axon terminals and transported along the axon's cytoskeleton into the neuron body.

Brainbow is a new method of tracing neuron connections. Brainbow uses a color labeling method based on the differential expression of several fluorescent proteins in the brains of genetically modified mice. Fluorescent proteins can mark individual neurons with one of over 100 distinct colors. The labeling of individual neurons with a distinguishable color allows tracing of their projections and reconstruction of their structure in small blocks of tissue (*Figure 7-5*).

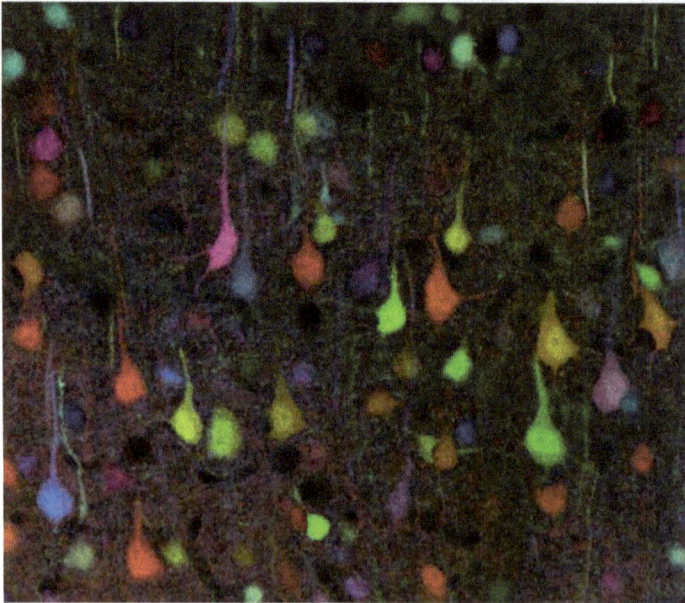

Figure 7-5: Mouse neurons labeled with fluorescent tags, Brainbow method. Photomicrograph: ©Stephen J Smith, source manuscript Pub-Med Central PMC2693015

LINKING ANATOMY TO PURPOSE

LOCAL BRAIN INJURY

Many theories stemming from the work of anatomists like Ramón y Cajal, Golgi and Brodmann speculate that variation in neuron structure, neuron location in the brain and neuron connectivity regulate distinct characteristics of brain performance.

Confirmation that intellectual pursuits align with discrete regions of the human brain emerged from examination of patients in the clinic who suffered local brain damage. For example, damage to specific brain areas is able to cause a general loss of understanding, deficits in the ability to use language or a lack of ability to control of emotion. Injury to the hippocampus affects the formation and duration of memories of life's experiences causing various degrees of amnesia.

While patients with brain injuries contribute a great deal to understanding of brain mechanisms, it should not be assumed all aspects of a missing capability are managed solely by the area with obvious damage. Normal activity requires dynamic interactions between several brain regions. A multistep process can be interrupted by removing or restricting a single step. Information carried by neuron circuits is splintered into many small parts as it progresses through the brain. Multiple bits of information about a single event distribute across multiple brain regions. Why this complex system is an advantage is not known.

EVENT-RELATED POTENTIALS

Modern human brain studies apply non-invasive techniques to explore information handling by neuron circuits. Four non-invasive neuroimaging techniques dominate this research. First, there is Event-Related Potentials (ERPs), part of an electroencephalogram (EEG). With ERPs the electrical activity measured by scalp electrodes is time locked to presentation of a sensory stimulus, a picture or spoken word. ERPs are exceptionally sensitive to voltage changes due to neuron activity in the cerebral cortex. This method is often used to study the regions of the cerebral cortex dedicated to learning language.

MAGNETIC RESONANCE IMAGING

A second form of brain imaging used for human studies is a variation of *magnetic resonance imaging* (MRI) named functional MRI (fMRI). A basic MRI scan produces a static anatomical picture of large brain structures in virtual slices of living tissue. The basic MRI serves to identify large brain structures associated with neuron activity measured by fMRI. fMRI is often employed for research and MRI is used for routine clinical evaluation of brain pathology.

Resonance in physics is the tendency of a system to oscillate with greater amplitude at certain frequencies. Most medical applications of MRI use a strong magnet to set up an efficient resonant frequency. As the resonance of magnetically excited hydrogen atoms in the brain decays, a radio frequency signal is emitted.

Depending upon timing of the resonance induced by the magnet, and of the radio frequency signal measurements, images are classified as T1 and T2. T1 and T2 acquired images look different. Tissue areas with high water content, a lot of hydrogen atoms, appear black in T1 images (*Figure 7-10*) and white in T2 images (*Figure 7-6*). Compare the area of the ventricles filled with cerebrospinal fluid in the two figures.

In T1 images white matter is light gray, gray matter is gray and cerebrospinal fluid appears black. T2 images show cerebrospinal fluid and edema in the tissue as white areas (*Figure 7-6*). Both white matter and gray matter appear an almost uniform gray in a T2 image.

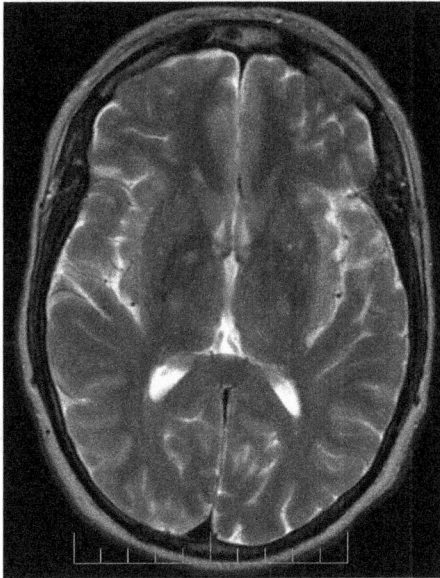

Figure 7-6: Standard MRI T2-weighted axial image of the brain showing cerebrospinal fluid as bright white. Image: ©Afiller

FUNCTIONAL MAGNETIC RESONANCE IMAGING

Functional MRI (fMRI) is similar in procedure to MRI, but it measures magnetization of oxygen-rich and oxygen-poor blood rather than magnetization of hydrogen in molecules. Oxygen is carried by hemoglobin molecules in red blood cells. Oxygen-rich hemoglobin is resistant to magnetism. In comparison, oxygen-poor hemoglobin is more magnetic. As one form of hemoglobin displaces the other a difference in the fMRI signal can be detected. The result of blood oxygen measurements is presented as a color-coded fMRI image (*Figure 7-7*).

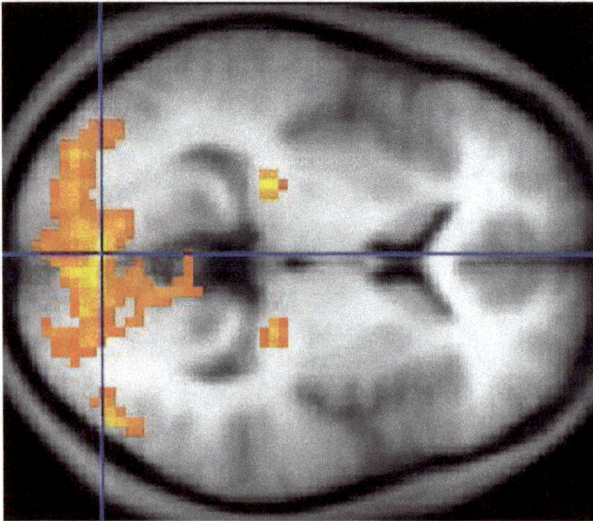

Figure 7-7: An fMRI image with yellow/orange areas showing brain areas with increased blood flow. Image: ©OpenStax College

Neuron activity requires a large amount of energy. Creation of energy rich molecules from glucose uses a lot of oxygen extracted from the hemoglobin of red blood cells. Astrocytes determine the rate of local brain blood flow. When neurons increase their signaling activity, surrounding astrocytes signal for increased local blood flow to insure an adequate oxygen supply. Oxygen rich blood displaces oxygen-depleted blood in about 2 seconds.

The slow dynamics of the blood flow response, seconds, compared to neuron response in milliseconds means a signal detected by fMRI is a summation of neuron activity in small tissue areas. The purpose of these kind of data is to correlate brain neuron activity with a task performed by the tested subject. Multiple brain areas in use during recognition of objects, emotional states or listening to speech can be evaluated for their degree of association with fMRI. fMRI produces data sets from up to 100,000 locations at the same time. These data provide three-dimensional maps with high spatial resolution of areas working together across the whole brain.

Magnetoencephalography

A third type of non-invasive imaging used in brain research is *magnetoencephalography* (MEG). Breaking this long scientific term into its root parts produces the approximate translation 'mapping magnetic fields in the head'. The MEG signal measures neuron current rather than the oxygenation state of hemoglobin in brain blood vessels. MEG records magnetic fields produced by currents flowing through brain circuits using sophisticated sensors named magnetometers. It allows precise localization of the neuron currents responsible for the magnetic fields. MEG resolves events with a precision of 10 milliseconds, or less. Magnetometers reside in a helmet placed over the subject's head (*Figure 7-8*).

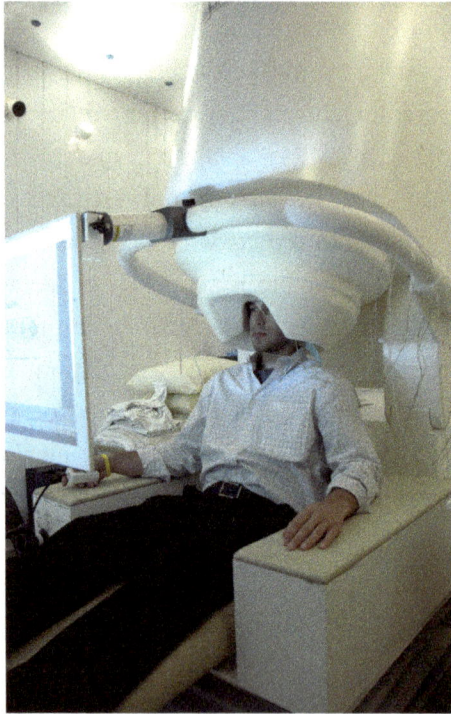

Figure 7-8: Person undergoing a MEG. Picture is from United States National Institute of Mental Health. This image is in the public domain in the United States.

Sometimes MEG is combined with fMRI when mapping a response to an experimental condition. The degree of agreement between data collected with these two techniques varies depending upon the complexity of the neuron circuitry in brain areas responding to the stimulus.

FUNCTIONAL NEAR INFRARED SPECTROSCOPY

The fourth method, functional near infrared spectroscopy (fNIRS), takes advantage of the transparency of skin, bone and brain tissue to light in the near infra-red light spectrum, 700-900 nanometers. Most of the thermal radiation emitted by objects near ambient temperature is in the infra-red spectrum. This light energy is invisible to the unaided human eye. Data obtained with fNIRS are similar to data produced by fMRI because they also measure changes in blood oxygenation of tissue.

An advantage fNIRS is it requires a simple helmet device and is therefore far less expensive than fMRI. A disadvantage of fNIRS is near infrared light cannot reach brain areas deeper than about 4 centimeters. Its use in humans is limited to studies of the cerebral cortical neurons. Another disadvantage of fNIRS is its lack of anatomical reference points for the observed changes in blood oxygenation.

HUMAN MEMORY

CATEGORIES OF MEMORY

Memory molds consciousness, making each mind unique. Patients who suffered brain damage to their hippocampus provided the first definitive knowledge that human memory is stored in, and recalled from separate brain regions. These patients lost their ability to form and recall particular types of memory, while retaining other types of memory. Individual losses depended upon the precise location of the injury.

Modern non-invasive brain imaging studies are producing volumes of data to define neuron circuits dedicated to storage and recall of human memory. There are multiple forms of human memory and different neuron pathways support each type. In brief, the major divisions of memory are *long-term memory* and *short-term memory*.

Most systems for cataloging the defining characteristics of human memory originated in the field of psychology. Psychologists report long-term memory can be divided into two separate categories, *explicit/declarative memory* and *implicit/procedural memory*. Short-term memory, also known as *working memory*, is more complex than long-term memory.

EXPLICIT/DECLARATIVE MEMORY

Explicit memory includes memory available at will. Explicit memories fit into three categories: *episodic memory, semantic memory* and *autobiographical memory*. Episodic memory recalls specific events, situations and personal experience. For example, an episodic memory may be of a person, of a person's name or the place and time spent with a particular person.

In contrast, semantic memory is limited to facts like the number of hours in a day or the meaning of a word. Autobiographical memory is similar to episodic memory but is restricted to those events in a person's own life history. Memory of events corresponding to a person's college graduation or getting married is autobiographical memory.

IMPLICIT/PROCEDURAL MEMORY

Implicit memory, unlike explicit memory, is below the level of conscious awareness. Implicit memory is an unintentional memory formed as an experience is repeated. Implicit memory guides performance of a practiced activity. It is the automatic form of memory used when reading this book. Attention to the details of how to read is not required because an implicit memory of reading is already present. In another example, implicit memory recognizes that music belongs to particular categories like classical, country, blues or rock.

WORKING/SHORT-TERM MEMORY

Working memory is not part of long-term memory. Rather working memory is a dynamic form of memory combining many things from past learning together with present experiences. It draws on previous experiences to manage the present.

Working memory also includes resisting distraction when processing information for storage as long-term memory. For example, when students hear instructions for how to proceed with an exam, working memory stores details of the exam process in long-term memory and at the same time formulates questions to ask about the directions.

Neuron memory circuits

Examination of neuron circuits that manage memory is an active area of research in neuroscience. However, few central concepts about memory formation and memory recall circuits enjoy general agreement. Studies of episodic memory are producing data best accepted.

Episodic memories are coded first by neurons of the hippocampus (*Figure 7-2*). Coded information travels from the hippocampus to an area of the temporal lobe of the brain named the *entorhinal cortex* (*Figure 7-9*).

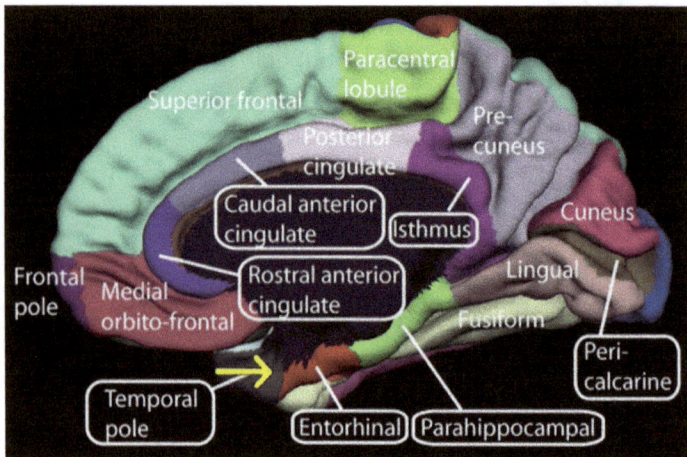

Figure 7-9: Medial surface of the cerebral cortex showing location of the entorhinal cortex and parahippocampal cortex of the right hemisphere. Illustration: Hagmann P, et al. 2009, PMCID: PMC2443193

The entorhinal cortex is the primary interface between the hippocampus and the neo-cortex where memory is stored. The entorhinal cortex is critical for long-term memory formation, memory consolidation and memory optimization by sleep. The neurons of the adjacent nuclei that are collectively known as the amygdala add emotion to remembered events.

Long-term episodic memories fragment and scatter over a network of neurons in diverse brain areas including multiple locations in the neo-cortex. The individual components of a single event move to separate locations for storage. Visual representation, auditory representation and the emotional component of an event may all map to different brain areas. To recall an episodic memory, the brain must reassemble the components.

The act of retrieving an episodic memory into working memory reshapes pieces of the older memory into a new episodic memory. Old memories become updated to include new experience with each recall. In this way the brain integrates what is already understood to be true with everything new it learns. Because of this updating tendency of the human mind, eyewitness descriptions of past experiences are unreliable.

Studies using fMRI during episodic memory formation followed by recall of the memory discovered remembering reactivates the same neuron circuitry as the original experience. There is general agreement, at least for episodic memory, that recall follows the same neuron path through the brain as the actual experience, but in reverse. However, the number of details remembered about a previous event depends upon the amount of attention focused upon the event when it occurred.

Unlike episodic memory, the main neuron circuit controlling implicit/procedural memory is a loop connecting various parts of the neo-cortex, the cerebellum, the spinal cord, the thalamus and the basal ganglia (*Figure 7-10*).

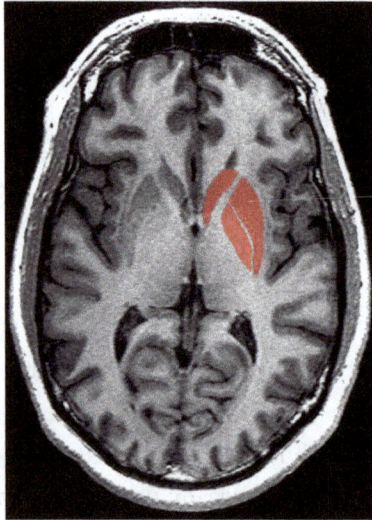

Figure 7-10: Tranverse section of the striatum, the largest of the basal ganglia nuclei, labeled in red on T1 MRI image. MRI: ©Lindsay Hanford, this image is released to the public domain.

Body movement is initiated by the neurons of the motor cortex in response to executive decisions made in the frontal cerebral cortex. The cerebellum fine-tunes the output of the motor neurons on their way to the spinal cord. The neurons of the motor cortex synapse on the motor neurons in the spinal cord sending a signal to the muscles. Sensory receptors detecting position of the body send information about muscle positions back to the spinal cord. The spinal cord forwards position information to the brain.

All feedback from outside the brain about task implementation enters the brain by way of the spinal neuron tracts to the thalamus. The thalamus relays feedback information from the muscles to the cerebral cortex, cerebellum and basal ganglia. The basal ganglia, a group of brain nuclei deep beneath the cortex, smooth out voluntary movement by coordinating the movement with conscious motivation. The basal ganglia accomplish this through their connections to the thalamus, the motor neurons of the cortex, the brainstem and other brain nuclei. The unconscious memory of many iterations of an activity becomes an implicit/procedural memory that is thought to be stored as a neuron template within the cerebellum.

ANATOMIC STRUCTURE OF MEMORY

NEURON CONNECTIONS

Memory formation in the human brain involves more than neuron current moving through stationary neuron circuits. Dynamic anatomic reorganization of neuron connections during memory formation is essential. Neuron circuit reorganization includes restructuring of neuron ensemble units, remodeling of neuron synapses and replacement of some neurons. These events are enhanced by sleep.

There is evidence for activity-dependent recruitment of neurons from reserve neuron pools to ensemble circuits. Reserve neuron pools exist in signaling ensembles of vertebrate brains from frogs to primates. The reserve pool neurons share input and output circuits with the core ensemble but express different neurotransmitters and exert different effects on common targets. Incoming stimuli affect both the core and reserve neuron pools. The flexible composition of ensembles allows various combinations of neurotransmitter output in response to incoming signals.

Neuron ensembles possess an additional mechanism for responding to variation in multiple sensory inputs of diverse intensity. The presumption asserting a neuron's neurotransmitter never changes is incorrect. For example, neuroscience once thought dopamine neurons always released dopamine and norepinephrine neurons always released norepinephrine. Today there is evidence that neurons can respond to fluctuating incoming electrical activity by changing the neurotransmitter they release and the neurotransmitter receptors they display.

White matter reorganization is also part of memory formation. Modification of axons in a circuit include alteration in the number of axons in a memory pathway, changes in the diameter of the axons and variation in the degree of dense packing, axon branching and the amount of axon myelination. The ability of neurons to adjust their properties during memory creation provides a dynamic range of options for optimizing memory construction.

PRESERVATION OF DENDRITIC SPINES

Memory programing begins at neuron synapses. Over time synapse structures enlarge or shrink depending upon the number of incoming action potentials. This suggests the amount of sensory input is correlated with the functional capacity of neuron ensembles. The more a pathway is used, the better it works. The process that allows synapses to become stronger or weaker is called neuroplasticity. A discussion of the plasticity of neuron synapses is presented in Chapter 5, *"Neuron synapses—Excitatory and Inhibitory."*

Studies during the 1970s predicted synaptic plasticity. In those studies, small groups of brain neurons were activated with an electrical current from tiny electrodes. Investigators reported a phenomenon they named *long-term potentiation* (LTP). Long-term potentiation is described as an increase in the ability of brain synapses to respond to neurotransmitter after receiving a rapid burst of signaling activity induced by external electrodes. The above-normal sensitivity of synapses to neurotransmitter persisted in these studies for minutes to many months.

The biologic mechanism supporting LTP remained unclear for a long time. Contemporary data demonstrate LTP requires a large increase in neuron cytoplasmic Ca^{++}, gene transcription in the neuron nucleus and new protein synthesis with dendritic spine enlargement. LTP affects both the presynaptic and postsynaptic compartments. These are the same changes observed in neurons routinely experiencing exceptional sensory input like those involved in memory formation. In fact, neurons expected to respond in LTP experiments reside in brain areas known to be associated with memory formation, the cortex, cerebellum, hippocampus and amygdala.

REPLACEMENT NEURONS

Neurogenesis, the birth of new neurons, in the adult brain is a modern discovery. During the 1990s a stem cell population capable of developing into new neurons was discovered in the mouse brain. Brain histology published in 1998 confirms the human hippocampus also retains an ability to generate new neurons throughout life. In the adult human brain neurogenesis appears to be restricted to the dentate gyrus (DG) region of the hippocampus (*Figure 7-2* and *Figure 7-11*).

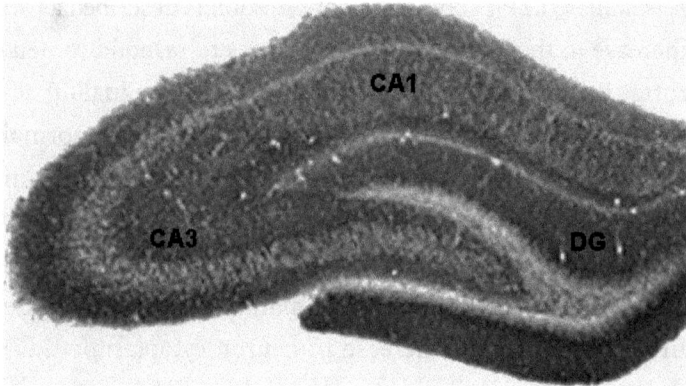

Figure 7-11: Regions of the hippocampus, DG = Dentate Gyrus region, CA = Cornu Ammonis regions of densely packed neurons similar to those of the neo cortex. Photomicrograph: Semiconscious, released to public domain by author

The dentate gyrus of the hippocampus contributes to formation of new episodic memories and recognition of environmental patterns. It also plays a role in emotional memory and recall of facts. Not long ago neurogenesis in the human hippocampus was sufficiently quantified to support the possibility that new neuron formation may be important to memory formation.

A 2013 study was the first to present an estimate of neuron turnover rates in the adult human hippocampus. The investigators measured the concentration of radioactive carbon (^{14}C) in DNA using postmortem brain tissue. The strategy took advantage of elevated ^{14}C in the atmosphere created by above-ground nuclear testing between 1955 and 1963. ^{14}C in the atmosphere reacted with oxygen to form $^{14}CO_2$ that entered the human food chain through plant photosynthesis of glucose. Once distributed through the body $^{14}CO_2$ created a date mark as it was incorporated into the nuclear DNA of new cells.

The 2013 study confirmed the 1998 histology study showing neurogenesis in the human hippocampus is restricted to the dentate gyrus. Because ^{14}C is radioactive and emits a low energy beta particle with a half-life of 5,700 ±40 years, these investigators were able to determine the rate of neuron turnover in the dentate gyrus. They concluded the vast majority of human dentate gyrus neurons are replaced over a lifetime. Their mathematical models predict a median turnover rate of 1.75% per year, corresponding to about 700 new neurons per day in each hippocampus, right and left hemisphere. Neurogenesis continues without decline into at least the 5th decade of life based upon the oldest individuals in the study.

The half-life of replacement neurons in the adult human dentate gyrus is about 7.1 years. This is 10 times shorter than the half-life of non-renewing neurons in the remainder of the hippocampus. Studies in rodents, where dentate gyrus neurons also turnover, indicate adult-born hippocampus neurons demonstrate enhanced synaptic plasticity and, therefore, a greater influence on hippocampus memory formation.

A two part model proposes pattern separation, which insures similar experiences persist as distinct memories, requires replacement hippocampus neurons for coding. The second component of this model maintains older neurons of the hippocampus, outside the dentate gyrus, are devoted to pattern completion. Pattern completion is essential for reconstruction of a stored memory based on partial clues.

SLEEP REINFORCEMENT OF NEURON CONNECTIONS

Many observational studies of human learning conclude there is a positive effect of sleep on the consolidation of recent events into long-term memory. However, empirical data are difficult to collect and proposed neuron mechanisms remain controversial. Studies investigating neuron synapse adjustments when learning new skills report sleep deprivation inhibits revision of dendritic spine synapses.

Imaging studies of the hippocampus report neurons there continue to be active during all phases of sleep. Neuron ensemble signaling patterns suggestive of episodic memory formation are also detected during sleep. For mammals, EEG wave forms during sleep and wakefulness are well established and can be associated with periods of specific types of memory formation.

Based on EEG wave forms, sleep is divided into two forms, rapid eye movement (REM) and non-REM sleep. The American Academy of Sleep Medicine further divides non-REM sleep into three stages N1, N2 and N3. Sleep stages follow each other in a sleep cycle. Sleep cycles last about 75 to 100 minutes (*Figure 7-12*).

Sleep Phases

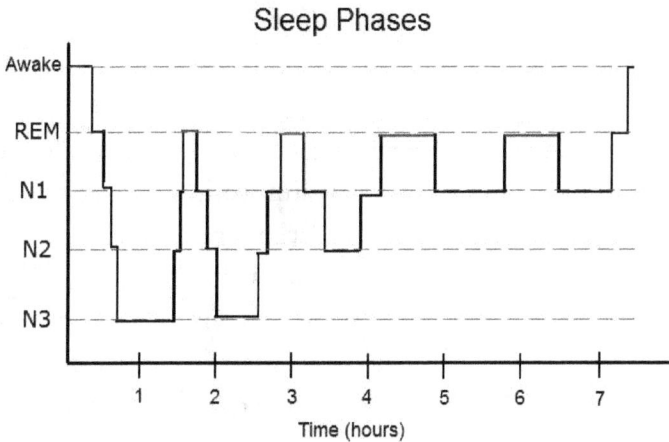

Figure 7-12: Graph showing the passage through the four principle phases of sleep over the course of a night. Portions marked in red indicate REM sleep. Illustration: ©Kernsters

Sleep cycles repeat during the period of sleep. The initial cycle includes a longer period of N3, slow-wave sleep, than subsequent cycles. N1 is the stage between wakefulness and sleep. The EEG pattern is rhythmic high amplitude, low frequency waves named alpha waves. In the N2 stage, abrupt activity described as sleep spindles interrupts the alpha waves. In slow-wave sleep, stage N3 the EEG waves reach their highest amplitude and lowest frequency.

The pattern of EEG waves during N3 is known as delta waves.

During REM most muscles are paralyzed and the EEG pattern is to a great extent like the EEG pattern when awake. Dreams occur during REM sleep. At the beginning of sleep, most of first sleep cycle is spent in the slow-wave N3 stage. As the cycles repeat, the amount of time spent in N3 stage decreases and the amount of time in REM sleep increases.

Imaging studies during slow-wave N3 sleep detect spontaneous reactivation of neuron pathways used to encode and store recent explicit/declarative memories. Reactivation of neuron tracts during this stage of sleep improves later recall of episodic and semantic memories. Spontaneous reactivation of neuron pathways associated with memory consolidation is also present during REM sleep.

REM sleep replay of the initial memory circuit pattern may last seconds or minutes. There is significant correlation of ensemble firing patterns between the conscious behavior and those of REM sleep.

Yet, during REM sleep the actual sensory input needed to stimulate the ensemble firing pattern is no longer present. Behavioral studies demonstrate a positive effect of REM sleep on implicit/procedural learning absent feedback from the paralyzed muscles. The effect of REM sleep on explicit/declarative memory is an area of ongoing research.

LEARNING LANGUAGE

NATIVE LANGUAGE

On average, hippocampus maturity is reached at about age 3-4 and coincides with the age when an infant acquires the ability to fit words together in a native language. This is also the age when people develop the earliest of the long-term memories revisited later in life. This convergence of developmental events implies language is needed for episodic memory formation.

The ability of humans to learn, communicate with and think in a language fascinates both neuroscientists and psychologists. The advent of non-invasive brain imaging techniques expanded knowledge of particular brain areas and neuron circuitry essential for learning language. Language acquisition studies include infants and young children, because they are superior learners of language compared to adults.

Remarkable data are accumulating describing establishment of a person's native language and the neuron circuits supporting language in general. Exposure of infants to language alone, like hearing a recording of a human voice, appears inadequate. To trigger learning from exposure to language there must be a social interaction between the infant and another human being. Critical aspects of the social components remain unclear.

The infant brain is wired for receiving auditory input prior to birth. Data indicate infants recognize their mother's voice at birth. They begin learning the sounds of their caregivers' language during the first 12 months of life. Their implicit memory comes to recognize that sounds fall into patterns with regularities in the verbal flow. The patterned sounds become words occupying grammatical blueprints (*Figure 7-13*).

Figure 7-13: Young mother talking with her baby. Photo: ©Kletr

Vocabulary expands fast around 18 months of age. Fitting words together into a native language begins between 18 and 36 months. Speaking the native language with fluency is typically mastered by age 8.

Among the theories of why adults experience greater difficulty learning a new language is the idea of interference. Neuron circuits supporting patterns of an established language may hinder recognition of the new language's patterns. Additional languages can be learned at any age but fluency of a native speaker is seldom attained if the new language is acquired after puberty.

How Small Children Are Studied

Event-Related Potentials (ERPs) reveal language processing in little children (*Figure 7-14*). ERPs measure the parts of an electroencephalogram (EEG) corresponding in time to a specific form of sensory input, in this case speech. ERP/EEG is inexpensive in comparison to other techniques. There is a good association between exposure to a stimulus and measurement of the responding neuron activity. Being noiseless is an additional advantage for auditory studies. Its major drawback is its sensitivity to movement.

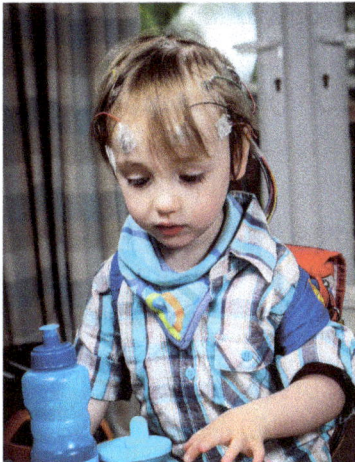

Figure 7-14: Two year old boy with 24 hour EEG electrodes attached. Photo: ©Steve Buckley

Magnetoencephalography (MEG) another noiseless technique is used for language acquisition studies of children up to 2 years of age. MEG's advantage is its ability to detect the precise location of the neuron circuits activated by a stimulus. Movement does not affect MEG data because the helmet contains head tracking sensors. MEG allows tracing of neuron activity in both young children and adults. It can be used to compare neuron activity fundamental to learning a native language with circuits activated during adult learning of a second language. Its major disadvantage is high cost.

Near Infrared Spectroscopy (NIRS) also provides good spatial resolution of the brain areas with increased neuron activity when a specific sensory stimulus is presented. The helmet is small and easy to fit to an infant and the procedure is noiseless. Because NIRS is sensitive to movement, studies are often completed while an infant is asleep.

Functional magnetic resonance imaging (fMRI) is used most often for adult studies. Its spatial resolution of neuron response is excellent because it includes anatomic landmarks. However, fMRI is sensitive to movement and is noisy. Its noise level limits its value in studies of small children because sound protection is necessary. fMRI is also expensive to use in research protocols.

LANGUAGE CIRCUITS

A mere 600 consonants and 200 vowels are used to create all of the world's languages. The English language uses a set of about 40 non-identical sounds made up of consonants and vowels referred to as *phonemes*. The dictionary defines phoneme as the smallest speech unit in a language capable of conveying a distinction in meaning. Languages vary to a great extent in the number of phonemes they use.

Babies begin learning language by selecting and figuring out the meaning of the sounds in the speech they hear. If an infant's parents use two different native languages, the infant is able to learn the rules for both languages at the same time. It is improbable this occurs because the infant brain stores information about the statistical patterns of the languages heard. Rather, multiple memory pathways induced by socialization play yet undocumented roles attaching meaning to the sounds of language.

The auditory and the speech cortex are located at Brodmann areas 22, 41, 42, 44 and 45. In humans damage to the primary auditory cortex causes only a loss of awareness of sound. Reflexive reaction to sound remains. The auditory and speech cortical areas are only some parts of the brain involved in learning language.

Permanent loss of language due to brain injury requires failure of specific subcortical areas as well. The basal ganglia, and pathways to and from these brain nuclei, play a significant role. Comprehending the meaning communicated by the order of words requires activity in the basal ganglia circuits to the prefrontal cortex.

Investigators puzzled for years over how the brain extracts and encodes speech from a noisy background. Linguists Roman Jakobson, Nikolai Trubetzkoy and Sergej

Karcevskij in a 1928 paper classified phonemes, sounds used in speech, into categories by how air is moved through an oral cavity when the phoneme is spoken. The three criteria used were 1) where air is compressed in the passage on its way out, 2) vibration or no vibration of the vocal cords 3) how air is released. They demonstrated phonemes could be assigned to groups based upon these vocalization characteristics.

Data obtained by recent studies of the language centers in the human brain support Jacobson's grouping of phonemes. Studies recording direct cortical activity in humans as part of their clinical evaluation for epilepsy surgery discovered the majority of the English speech-activated locations in the auditory cortex respond to a particular phoneme group.

This result is in contrast to an earlier expectation that each phoneme would activate a different location on the auditory cortex. Segregation of auditory cortex response elements by groups of phonemes based upon the mechanics of speaking implies speech and language may develop together.

Some investigators hypothesize the difficulty experienced by adults in learning a new language is due to circuitry constructed in infancy to support phoneme groups of the native language. Overlap of phoneme groups between languages could trigger inappropriate associations in the adult learner. It may seem to the adult learner that certain words represent something different than their true sense.

Language is managed in a different way in the brain's left and right hemisphere. The left hemisphere is the primary language manager for the majority of people. The right hemisphere is less involved, but communication between the hemispheres is critical for comprehension of ambiguous words like 'lead', which can either mean 'a metal substance' or to 'act as a guide'.

Results of recent imaging studies discourage theories of complete dominance by a single hemisphere of any particular property of language. Both MEG and fMRI studies report the involvement of both hemispheres in many aspects of language processing. Dominance is now discussed in terms of better performance of individual tasks by one hemisphere or the other.

Large studies continue worldwide to improve tools available for investigation and modeling of brain circuits. Brain circuits appear a good deal more complex and more fluid than ever imagined. It is also becoming clear that neuron circuitry is influenced by the non-neuron cells of the brain. Both astrocyte networks and patterns of myelination by oligodendrocytes impact the output of neuron ensembles.

Neuron ensemble signaling pools and neurons capable of changing their neurotransmitter in response to environmental fluctuation add an additional challenge. Uncovering the mysteries of the closed world of the human brain will require co-operative efforts by mathematicians, physicists, chemists, engineers and neuroscientists. There is much left to discover, but new studies each year open a clearer picture of how the human brain works.

SUMMARY CHAPTER 7

- Evidence for how memory is formed is the closest science comes to understanding consciousness, pleasure, anger, happiness, love and hate
- The 2-4 millimeter deep layer of neurons composing the human cortex is critical to language, memory, attention, awareness, thought and decision making
- Neuron information coding in the brain is carried out by populations of neurons working together in ensembles
- The connectome projects are mapping all neuron connections throughout the human brain
- EEGs measure large fluctuation in the brain field potentials caused by neuron ensembles firing action potentials in synchrony
- Four forms of non-invasive imaging for brain activity studies include Event-Related Potentials (ERP part of an EEG), functional magnetic resonance imaging (fMRI), magnetoencephalography (MEG) and functional near infrared spectroscopy (fNIRS)
- Imaging studies discovered sensory information from a single event is divided into many small pieces and is distributed widely across brain regions

- Storage and retrieval of episodic memories by the brain use the same neuron circuits, but the direction of current flow for retrieval is the reverse of that for storage
- Memory divides into two types, conscious explicit/declarative memory of facts and events and unconscious implicit/procedural memory of a practiced skill
- The neurons of the hippocampus code memory into signals that activate the nearby entorhinal cortex, which in turn moves pieces of memory to the neo-cortex for storage
- A template for unconscious memory of a practiced activity is thought to be stored within the cerebellum
- An essential component of memory formation is anatomical reorganization of neuron synapses called neuroplasticity
- Neurogenesis in an adult human brain is confined to the dentate gyrus of the hippocampus
- Sleep produces a positive effect on memory formation
- Infants and young children are superior learners of language compared to adults
- Segregation of auditory response elements by phoneme groups aligned with the mechanics of speaking implies speech and language may develop together

[8]

When It All Goes Wrong —Alzheimer's Dementia

LIVING A LONG LIFE is highly desirable. Even though memory processing adjusts as the human brain ages certain people, including some who live past 100 years, retain their mental competence. Yet, many adults fear onset of dementia in general, and Alzheimer's disease in particular. According to the United States Centers for Disease Control and Prevention, Americans are two times more afraid of losing their mental capacity than their physical capacity when they grow older.

But, what exactly is Alzheimer's disease, and is it the same thing as dementia? Contemporary data indicate Alzheimer's disease is a gradual change in the brain that occurs over decades before mental dysfunction is noticed. People who develop Alzheimer's disease lose their ability to remember and reason in a progression that is slow, frustrating and ultimately fatal.

Gradually neurons die in the regions of the brain needed for memory of people and events. During the pre-symptomatic phase of Alzheimer's disease, surviving cells in the brain compensate for the loss of neurons until finally the number lost becomes too great. The long pre-symptomatic phase of Alzheimer's disease may be viewed as good news, because it provides a lengthy period when prevention strategies can be designed and tested.

Abundant loss of neurons, and a resulting decrease in brain volume that proceeds sequentially through specific regions associated with memory and reasoning, differentiates Alzheimer's disease from other forms of dementia. An estimated 44 million people worldwide lived with dementia in 2014. The number is expected to rise to 70 million by 2030. Alzheimer's disease is the most common type of dementia and it accounts for 60% to 80% of all cases. By 2050 the projected cost of caring for individuals with Alzheimer's disease could reach $1.2 trillion in the United States alone.

This chapter examines the power of brain imaging techniques to unravel the cellular behaviors that produce Alzheimer's disease. Advances in imaging technology, designed for medical diagnosis, open a new view of the transformations occurring in the human brain with age. New understanding of the physiology of the brain presents unexplored avenues for disease prevention.

ALZHEIMER'S BRAIN

Differentiation of Alzheimer's disease from other forms of dementia is accomplished by examination of brain tissue after death. Postmortem evaluation of Alzheimer's brain reveals shrinkage in the hippocampus and in areas of the temporal lobe, parietal lobe and frontal lobe of the brain (*Figure 8-1*). These brain areas manage language, memory, emotion and the ability to reason.

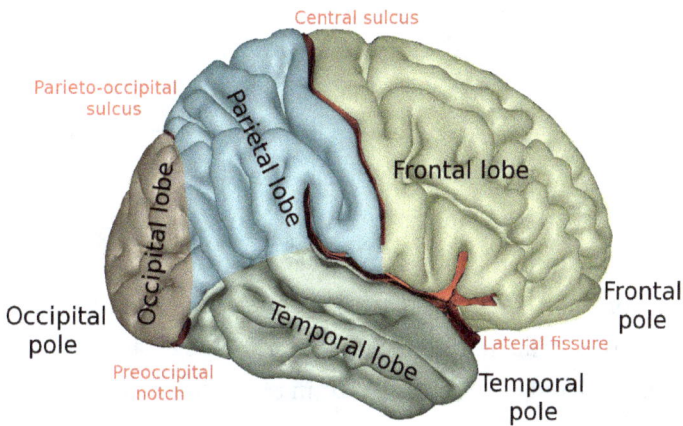

Figure 8-1: Lobes of the human brain showing main sulci and boundaries. Illustration: ©Sebastian023

The classical theory of Alzheimer's disease assumes deposits of *amyloid-β*, a small but aggregated protein, in some unidentified manner initiate a sequence of events leading to death of brain neurons. In turn, neuron death causes shrinkage of the particular brain areas believed to govern memory and emotions. In support of this theory, brains of deceased Alzheimer's patients display large aggregates of amyloid-β in areas lacking a normal quantity of neurons.

Postmortem histology of Alzheimer's brain detects signs of damage within surviving neurons and large amyloid-β deposits surrounding them. The most obvious internal neuron damage is the presence of a fibrous deposit described as *tangled tau*. In healthy brain, tau is a soluble protein within axons and dendrites and it would not be observed by routine histologic techniques. The anatomy of a neuron is illustrated in *Figure 4-1* in Chapter 4, *"Neurons— How They Make Electricity."*

The intellectual impairment observed in Alzheimer's disease correlates well with the number of neurons displaying tau tangles and the decreased size of affected brain regions. The presence of tangled tau in the neurons of Alzheimer's brains is assumed to signify a compromised cellular transport system. Soluble tau regulates assembly and disassembly of train track-like structures that transfer material from one section of a neuron to another. The damaged neuron transport system probably contributes to the shrunken appearance of neurons in Alzheimer's brain.

ALZHEIMER'S THERAPIES

Until a short time ago, it was hoped removal of amyloid-β plaque would halt disease progression for patients assumed to be experiencing Alzheimer's dementia. Clinical treatment trials involved delivery of antibody drugs to the brain. The antibodies were expected to assist the brain's innate immune system, the microglia, destroy the amyloid-β deposits.

Unfortunately treating patients with mild to moderate dementia, who also presented a genetic predisposition to Alzheimer's disease, with an antibody against amyloid-β failed to delay progress of their disease. The antibody employed in the clinical trials passed through the blood brain barrier and cleared amyloid-β deposits, but participants showed no improvement and their disease progressed.

Lack of positive results in clinical trials targeting amyloid-β plaque led some investigators to question whether its presence in Alzheimer's is a cause or symptom of the disease. GBI Research (Global Business Intelligence Research) reported in March, 2015 that 583 new drugs are in development for Alzheimer's disease. Most of these are in the discovery and pre-clinical testing phase.

New drug candidates favor different mechanisms of action than the earlier treatments, because therapies for Alzheimer's disease since 2006 failed at a rate of 95% across all three phases of clinical testing. Innovative new drug designs feature the latest discoveries about the aging brain.

PRE-SYMPTOMATIC ALZHEIMER'S

INVESTIGATIVE TOOLS

Combinations of non-invasive and invasive imaging techniques investigate accumulation of amyloid-β and tau in the brains of people at high risk for developing Alzheimer's disease. Evidence implies progression of Alzheimer's disease begins years to decades before symptoms of intellectual failure surface. Sequential, progressive change in amyloid-β deposits and tau's presence in brain fluid proceeds for an average of 10 years before even minimal mental impairment is experienced.

Pre-symptomatic Alzheimer's disease is investigated using non-invasive methods including electroencephalogram (EEG), magnetic resonance imaging (MRI), functional magnetic resonance imaging (fMRI) and magnetoencephalography (MEG). Specifics of these methods are described in Chapter 7, *"Brain's Infrastructure for Memory and Language."*

Positron emission tomography (PET), another imaging technique for studying pre-symptomatic Alzheimer's disease, is invasive because it requires injection of radioactive molecules. A radioactive tracer must be incorporated into a biological molecule capable of crossing the blood brain barrier. In the brain, the radioactive molecule emits a signal that can be measured after binding to a brain substance or after its modification by a brain enzyme.

The biologic tracer molecules chosen for PET studies of Alzheimer's disease include fluorodeoxyglucose (FDG), an analogue of glucose, and Pittsburg compound B (PiB), a molecule that binds to amyloid-β plaques. With FDG, Alzheimer's investigators look for brain areas with decreased utilization of glucose (*Figure 8-2*). Decreased consumption of glucose is interpreted as decreased neuron electrical signaling activity. PET data can be overlaid on MRI structural data as in *Figure 8-2* for a precise location of areas with low neuron activity.

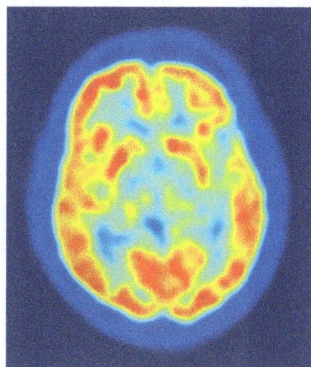

Figure 8-2: A transaxial slice of the brain of a patient taken with positron emission tomography (PET). Red areas show a high amount of ^{18}F-FDG tracer accumulation and blue areas low to no tracer accumulation. Image: ©Jens Maus

PET and fMRI remain the imaging techniques most often employed to study Alzheimer's disease. In September, 2014 a search of PubMed, the United States National Library of Medicine at the National Center for Biological Information, produced for the search terms "Alzheimer's + fMRI + human" a list of 5,165 papers. The earliest paper on the list was published in 1983. A similar search with the terms "Alzheimer's + PET + human" produced a list of 1,981 papers, and again the earliest paper on the list was published in 1983.

Both fMRI and PET use indirect measurements for evaluation of neuron signaling activity. fMRI measures flow of oxygenated hemoglobin into small areas of brain and PET measures glucose consumption. MEG, unlike fMRI and PET, measures actual neuron electrical activity and, though expensive, may be used to a greater extent in future studies of Alzheimer's disease. A PubMed search with the terms "Alzheimer's + MEG + human" produced a list of 111 studies since 1996.

BIOMARKERS OF ALZHEIMER'S DISEASE

Abnormal levels of soluble tau and soluble amyloid-β in the cerebrospinal fluid precede mild cognitive decline by several years. Tau increases and amyloid-β decreases in the cerebrospinal fluid during this time. The decrease in amyloid-β in cerebrospinal fluid is thought to reflect a reduction in its normal rate of elimination from the brain.

The increased presence of soluble tau in the fluids bathing neurons and in cerebrospinal fluid prior to onset of Alzheimer's symptoms suggests abnormal regulation of tau. Tau's normal work is thought to be accomplished inside neurons. Because tau is also present inside the non-neuron cells of the brain, the glia, neurons cannot be verified as the sole source of elevated tau in brain fluids during the pre-symptomatic phase of Alzheimer's disease.

Presence of tau outside neurons and the tangled form of tau protein within neurons are not restricted to Alzheimer's disease. A number of other neurodegenerative diseases display irregularities in the properties of tau protein within neurons and within glia. Neurodegenerative diseases with tau anomalies are assigned the general name *tauopathies.* They include among others frontotemporal dementia with Parkinsonism linked to chromosome 17 (FTDP-17T), ganglioglioma and lead encephlaopathy.

All tauopathies include neuron death and loss of the structures neurons use to communicate with each other named synapses. The name *synapse* derives from a Greek term, *sunapsis* meaning <u>point of contact</u>. An in depth description of activities at neuron synapses is presented in Chapter 5, *"Neuron Synapses—Excitatory and Inhibitory."*

PET imaging with Pittsburg compound B (PiB) estimates a brain's insoluble amyloid-β load and indicates its distribution in various brain regions. PET PiB studies detect accumulation of amyloid-β deposits in the brain for many years prior to appearance of memory loss in the portion of the population at high risk for developing Alzheimer's disease.

Below normal consumption of glucose in the brain is another characteristic of all Alzheimer's patients. Diminished utilization of glucose in particular brain regions is considered an indicator of neuron loss. PET with ^{18}F-FDG detects low glucose metabolism in the posterior cingulate, parietal lobe, temporal lobe and prefrontal lobe of some people at risk for developing Alzheimer's disease.

People at risk because of variations in their genes, and who develop the disease, sometimes exhibit a decline in their brain's consumption of glucose 20-30 years prior to symptoms. Yet, not every person with predisposing genes develops Alzheimer's disease. And, only about 50% of people with Alzheimer's disease have genes that are identified as high risk variants.

The pattern of activity between neuron circuits in various brain regions is studied using fMRI and MEG. One theory based on fMRI and MEG data proposes multiple functional rearrangements of neuron circuit activity may postpone cognitive difficulties during the pre-symptomatic phase of Alzheimer's disease. Neuron circuitry rerouting of information is estimated to occur as early as 10 years before symptoms appear.

Amyloid-β and Tau Physiology

Amyloid-β Physiology

It seems reasonable to assume the brain would not possess an elaborate process for creating amyloid-β if it served no purpose. Amyloid-β as a single peptide chain is a flexible and unstructured protein. The normal concentration of soluble amyloid-β protein in the human brain and cerebrospinal fluid is about 10^{-9} Molar, which is 4.5 micrograms of amyloid-β per liter of fluid.

During sleep-wake cycles, amyloid-β production is enhanced during periods of increased neuron activity and is decreased when neuron activity is low. Theories from this and other data propose amyloid-β is part of a negative feedback loop controlling neuron signaling.

Amyloid-β is a small peptide cut from a larger protein, *amyloid precursor protein* (APP) that is anchored in neuron membranes. Amyloid-β excision and replenishment of amyloid precursor protein in the membrane increase with rapid neuron signaling. Production continues until the concentration of amyloid-β becomes high enough at the synapse to decrease the effectiveness of the neurotransmitter released during signaling events (*Figure 8-3*).

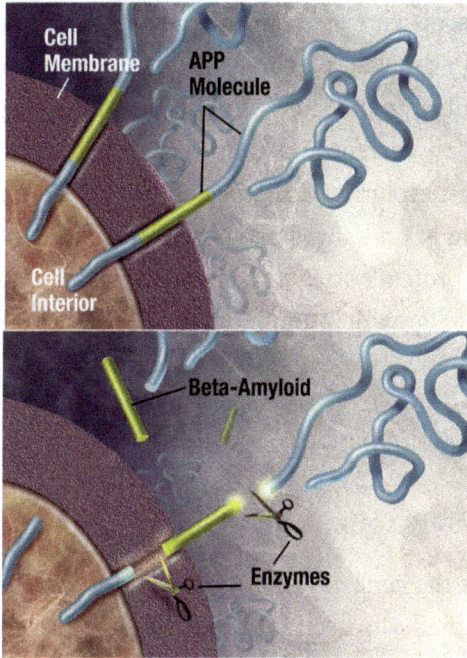

Figure 8-3: Amyloid precursor protein (APP) molecules in a cell membrane. The section of the APP molecules shown in green is amyloid-β. Illustration: National Institute of Aging, in the public domain in the United States.

A complex bell-shaped relationship between synaptic amyloid-β and neuron activity is proposed. The bell-shaped model hypothesizes intermediate concentrations of amyloid-β, 10^{-12} Molar to 10^{-9} Molar, encourage release of neurotransmitter. In contrast, levels below 10^{-12} Molar block neurotransmitter release. High levels of amyloid-β, above 10^{-9} Molar, also inhibit neuron signaling, but they accomplish it by blocking the effects of neurotransmitter rather than affecting its release from neurons. Similar bell-shaped activity relationships are common in biologic systems.

Less is known about the purpose of the larger segment of the amyloid precursor protein released by neuron enzymes during amyloid-β formation. Amyloid precursor protein is an ancient and conserved molecule throughout mammalian evolution. It occurs on the membrane of cells outside the brain as well as on neurons. Some of amyloid precursor protein's structural characteristics appear similar to those of growth promoting molecules like epidermal growth factor. Based upon its structure alone, the large piece released when amyloid-β is cut out may act as a growth factor in the brain.

Amyloid-β Plaque

As the concentration of amyloid-β increases to 10^{-6} Molar, it binds to itself and acquires a protein structure known as a beta sheet. The dramatic change in the configuration of amyloid-β from a floppy single strand peptide to a variety of multi-stranded, rigid structures encourages a theory maintaining each of amyloid-β's multiple forms participates in a unique brain function.

A contrasting theory proposes beta sheet amyloid structures may provide a simple and efficient way to remove excess functional amyloid-β by transforming it into a shape no longer useful at synapses. The reduction of amyloid-β in cerebrospinal fluid during pre-symptomatic Alzheimer's disease is interpreted failure of normal clearance of the molecule from brain rather than decreased production of the molecule by neurons.

The reasoning argues that amyloid-β would not accumulate as plaque if its production decreased. And, a slower than normal clearance rate of amyloid-β would encourage formation of the beta sheet proteins which aggregate into plaque. Proposed mechanisms for reduction of amyloid-β in cerebrospinal fluid include changes in its uptake by brain cells or a decrease in its transport across the blood brain barrier into the cerebral blood vessels.

Pinocytosis

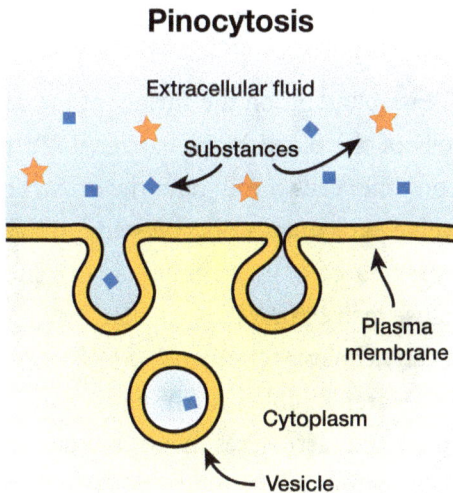

Figure 8-4: Pinocytosis is a form of cellular uptake. It uses areas of the cell membrane to grab small molecules from the surrounding fluid. Illustration: This work is released into the public domain worldwide by its creator Jacek FH

A variety of cells in healthy brain remove soluble single-strand amyloid-β. Surveying microglia, the neuron monitoring system of the brain, internalizes soluble amyloid-β by pinocytosis (*Figure 8-4*) and delivers it to lysosomes for degradation. Soluble, single peptide amyloid-β is sensitive to the many enzymes that as a matter of course digest proteins that are no longer needed.

Astrocyte glial cells and the endothelial cells of blood capillaries use a different mechanism than surveying microglia for removal of soluble amyloid-β. Both astrocytes and capillary endothelial cells participate in safekeeping of the blood brain barrier and in transfer of amyloid-β across the barrier. They use a membrane protein, *LDL receptor-related protein 1* (LRP1), to transport single-strand amyloid-β out of the brain.

LRP1 is abundant on the surface of the capillary endothelial cells of the blood brain barrier. LRP1 receptor is also found on other cells throughout the brain, because it is the major vehicle for uptake of cholesterol. Cholesterol is a critical component of brain membranes in general and of neuron synapse membranes in particular.

For fatty cholesterol to bind to LRP1 it must be surrounded by a water-soluble protein named *apolipoprotein E (ApoE)*. In the brain, astrocytes make cholesterol and package it in ApoE before secreting it for use by neurons and other brain cells. In contrast, soluble amyloid-β does not appear to be incorporated into ApoE packages before binding to LRP1. Rather, data indicate single-strand amyloid-β binds directly to LRP1 competing with ApoE for LRP1 binding sites.

The competition between amyloid-β and ApoE opens the possibility that a variant, stronger binding ApoE may prevent amyloid-β binding to LRP1. People who inherit genes coding particular variant forms of ApoE suffer a high risk of developing Alzheimer's disease. Possibly, specific variants of ApoE prevent efficient elimination of amyloid-β by LRP1.

Tau Physiology

Under normal circumstances, tau is a soluble protein within neurons. It provides for adequate spacing between track-like assemblies for moving material from place to place within neurons. Tau's efforts are controlled by an array of enzymes that add and remove phosphate groups at various positions. At least 30 sites on tau protein undergo modification by regulatory enzymes.

In addition to its well-documented role in neuron axons, tau is detected in the long dendrites at the other end of neurons. Its presence in axons, dendrites, dendritic spines and the neuron body suggests tau is involved in the ongoing transport of material between all parts of neurons.

Neurons often mature into long cells where proteins and molecules must travel long distances from their place of production to their destination. Feedback molecules from various parts of the neuron also travel along tau regulated assemblies to synchronize enzyme activity and gene transcription in the neuron body.

Tau protein lacks a necessary amino acid sequence for its direct passage through cell membranes. Yet soluble tau is found in normal cerebrospinal fluid. How and where along the neuron cell membrane tau may move into the neuron's bathing fluid, and from there into the cerebrospinal fluid, is unknown. Additional information about the fluids of the brain is presented in Chapter 3, *"Quality Control of Brain's Extracellular Fluids."*

Tau without phosphate groups attached is a flexible soluble protein lacking structure similar to single-strand amyloid-β. Recent studies of the non-phosphorylated tau molecule detect regions that assume a helical formation when near the lipids of the cell membrane. All proteins capable of passing through cell membranes assume a helical form within the membrane. Discovering whether this property is an important factor in tau's ability to move out of neurons requires further investigation.

Tau observed within neurons in postmortem Alzheimer's brain possesses a larger number of phosphate groups than required for its normal activities. The extra phosphates change tau from a flexible molecule to a helical filament with a negative electric charge. Pairs of helical tau filaments aggregate into the tau tangles visible within neurons.

Extracellular, soluble tau is elevated in the neuron synapses during Alzheimer's disease. Soluble tau within the Alzheimer's synapses acquires some phosphate groups but not so many that it precipitates out of solution.

NEURON DAMAGE AND LOSS

MEMORY CIRCUITS

Postmortem Alzheimer's brain shows a dramatic decrease in total volume in addition to amyloid-β deposits and tau tangles inside remaining neurons. The entire loss in size of the brain appears to be due to disappearance of neurons. The astrocyte cell population appears to remain stable and the microglia population expands.

Neuron loss is particularly evident in the hippocampus and cerebral cortex of the temporal, parietal and frontal lobes. Regional loss of volume is sequential and consistent with disease progression (*Figure 8-5*).

Brief periods of extreme loss of recall of events, situations and personal experiences provide an initial warning of Alzheimer's disease. The part of the brain that codes memory of events, episodic memory, is the hippocampus. The hippocampus and its adjacent brain areas are among the first to suffer damage by Alzheimer's disease.

The hippocampus receives sensory input from auditory, visual and olfactory cortical neurons. It also receives input from the neurons of the frontal cortex, the part of the brain in charge of making executive decisions. Hippocampus neurons send their memory codes to a portion of the temporal lobe named the entorhinal cortex. From there pieces of each memory move to other parts of the cerebral cortex for storage until they needed in the future. Neuron input from the neighboring amygdala to the hippocampus adds emotional texture to memories.

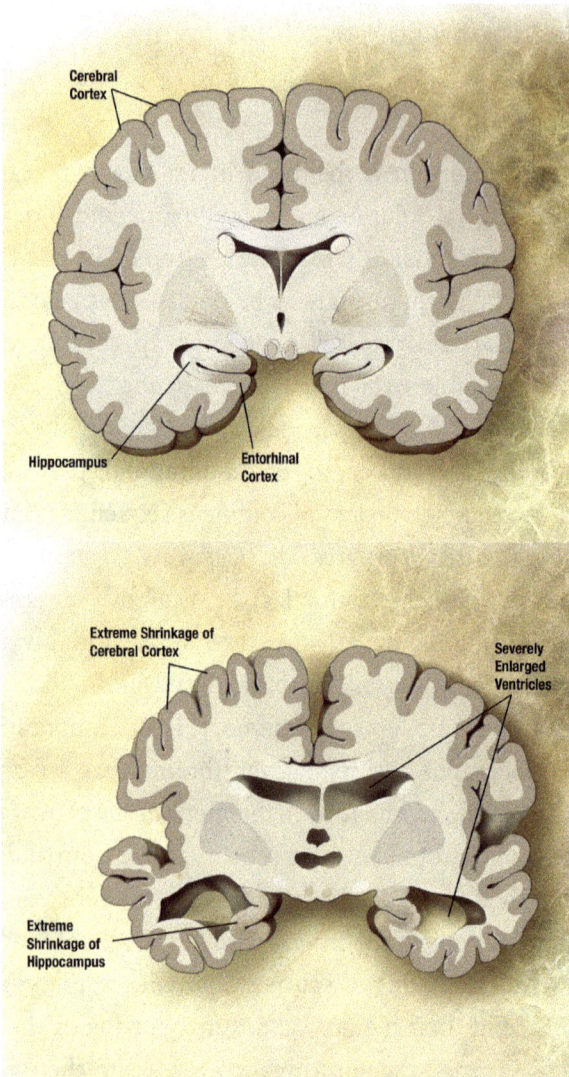

Figure 8-5: Alzheimer's disease brain comparison. The top is an illustration of normal human brain structure and the bottom is an illustration of a brain with Alzheimer's disease. ©Garrondo, released to the public domain

A large MRI study evaluated volume and shape change in areas of the medial temporal lobe of people at risk for developing Alzheimer's disease. MRI scans were acquired from 1995 to 2005. By the time the study published in 2014 participants were followed for 18 years. The goal of the study was to determine the sequence of affected brain areas during pre-symptomatic Alzheimer's disease.

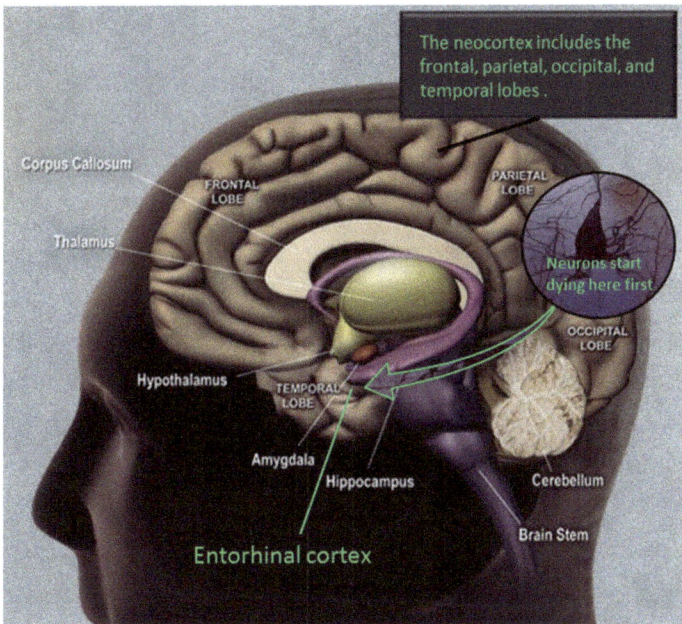

Figure 8-6: Areas of the brain that fail as Alzheimer's disease progresses. Illustration: ©7mike5000

In this MRI study, the entorhinal cortex shows significant volume change as soon as 8 to 10 years before the onset of symptoms. Atrophic changes in the entorhinal cortex align with the pre-symptomatic time points when amyloid-β first decreases in cerebrospinal fluid. The hippocampus shows atrophy 2 to 4 years prior to symptoms. And the amygdala displays significant changes 3 years prior to mild cognitive impairment. Atrophic regions correspond to brain areas postmortem with shrunken neurons, heavy deposits of tau tangles and amyloid-β plaque (*Figure 8-6*).

Glucose Metabolism

A decrease in glucose metabolism revealed by PET with ^{18}F-FDG, a radioactive glucose analog, aligns with onset of brain atrophy detected by MRI. Neurons depend almost exclusively upon glucose for energy, and their signaling procedures require about 20% of the body's total energy supply. PET data combined with fMRI data showing reduced blood volume in the brain areas affected by Alzheimer's disease suggest the disease initiating factor may be a failure in the brain's energy producing metabolic pathways. Complex models have been proposed for how neurons and astrocytes may share energy production. The details of the pathways are less important for this discussion than the fact that the metabolic reliance of neurons and astrocytes upon each other is complex.

If the metabolism theory of Alzheimer's disease is correct, then the metabolic defect may be an astrocyte defect, or a neuron defect or both. Glucose metabolism in brain requires a partnership between neurons and astrocytes. It entails a dynamic sharing of molecular conversion pathways between the two cell types. A detailed discussion of the normal dynamic relationship between neurons and astrocytes for energy conservation is presented in Chapter 6, *"Introduction to the Glia and Microglia—Meet the Stage Crew."*

ACTIVATION OF GLIA AND MICROGLIA

REACTIVE ASTROCYTES

Astrocytes tile the brain in a close ordered lattice. Based upon their morphology, three types of astrocytes reside in the brain. They are protoplasmic astrocytes, fibrous astrocytes and reactive astrocytes. *Protoplasmic astrocytes* surround neuron bodies. *Fibrous astrocytes* associate with neuron axons. *Reactive astrocytes* disperse throughout damaged brain.

When brain tissue experiences an injury, protoplasmic and fibrous astrocytes react by shifting their characteristics to those of reactive astrocytes. The astrocyte response to brain damage is termed reactive astrogliosis. Astrogliosis is a variable modification of astrocyte behavior depending upon the context of the damage. In the less severe form of astrogliosis, reactive astrocytes adjust and repair local brain structure without scar formation. When damage is severe, reactive astrocytes proliferate and form scars to wall off the damage.

A large number of reactive astrocytes inhabit brains with Alzheimer's disease. The shift from an interspersed population of supportive protoplasmic and fibrous astrocytes to a population where reactive astrocytes predominate begins in the initial stages of Alzheimer's disease. With increasing duration of disease, the portion of the astrocyte population in the reactive form increases.

Careful studies of the astrocyte population of Alzheimer's brains suggest the total population of brain astrocytes does not change during disease progression. Rather the percent of the population functioning as protoplasmic and fibrous astrocytes decreases while the percent functioning as reactive astrocytes increases.

ACTIVATED MICROGLIA

Two functional forms of microglia live in brain tissue, surveying microglia and activated microglia. During the switch from surveying form to the activated form, microglia changes its shape, proliferates and begins to express markers characteristic of the macrophages of the body's immune system. The activated form of microglia removes neurons with severe damage and engulfs amyloid-β plaque.

Of note, in Alzheimer's disease all dead neurons are missing from the brain. Dead and dying neurons are not identified in the tissue. In healthy brain surveying microglia monitors the well-being of neuron synapses. If a neuron is damaged surveying microglia attempts to repair it by secreting growth factors. However, if a neuron is damaged beyond repair surveying microglia transforms to activated microglia, finishes the kill and removes the debris.

In Alzheimer's disease it is unknown whether surveying microglia fails to repair damage to neurons caused by some yet unknown factor, or if surveying microglia plays a role in initiating neuron death before becoming activated and disposing of the remains.

Uptake of single-strand amyloid-β by surveying microglia to clear it from brain is different than the response of activated microglia to amyloid-β plaque. Activated microglia treats amyloid-β plaque as a foreign intruder. It attempts to surround and destroy the plaque. This is the same form of microglia that removes dying neurons. Because activated microglia responds to amyloid-β plaque as a foreign intruder, Alzheimer's disease is described as an *inflammatory disease*. This tends to cause confusion, because the inflammation of Alzheimer's disease exhibits different characteristics than classical *inflammation*.

Classical inflammation is a process involving cells of the body's immune system responding to the presence of a foreign substance. The classical inflammatory response dilates blood vessels and increases their permeability to water and cells. Water and immune cells then gain access to the foreign object. Heat, swelling, redness and pain develop in the affected tissue. Destruction or walling off of the foreign object is followed by healing in the tissue.

Classical inflammation does not occur in the brain unless the blood brain barrier is broken. The blood brain barrier of deceased Alzheimer's patients shows no functional loss beyond that normally expected because of age. Pathologists label Alzheimer's brain "degenerative" rather than "inflammatory" on postmortem evaluation. The term 'inflammation' when characterizing a brain displaying Alzheimer's disease refers to the presence of a large quantity of reactive astrocytes and activated microglia.

In Alzheimer's brain activated microglia surrounds dense amyloid-β plaques but appear unable to remove them. Like macrophages of classical inflammation, activated microglia discharges cytokines, chemicals that attract other cells to the areas of plaque invasion. The cytokines released by microglia draw reactive astrocytes to the plaque.

Reactive astrocytes internalize amyloid-β using their cell membrane proteins. Reactive astrocytes also encircle amyloid-β plaques in formations similar to the glial scars they create after brain trauma. Glial scar-like formation may create a barrier to separate amyloid-β plaques from healthy brain tissue.

Alzheimer's-like Brain without Dementia

Alzheimer's disease was originally described as a form of dementia accompanied by the presence in the brain of large amyloid-β plaques between neurons and tangled material within neurons. A feature lacking in histologic investigations of Alzheimer's disease up until publication of the Nun study in 1997 was a control group of brains from old people who remained mentally competent during their lifetime.

A similar but larger study, the Religious Orders Study, funded by the United States National Institute on Aging at Rush University Medical Center in Chicago will continue through June, 2016. By the time of completion, the study will have 22 years of clinical data on more than 1,000 participants and brain tissue from over 350 people. Nuns, priests and brothers from 31 Catholic orders in the United States are participating.

A surprising outcome of the controlled studies of age and Alzheimer's disease is the discovery of participants with intact intellectual function whose brains display Alzheimer's-like amyloid-β deposits and tau tangles at postmortem examination. This phenomenon appears in the results of both the Nun Study and the Religious Orders Study.

In these studies, 12% to 30% of participants without cognitive impairment at death displayed substantial amounts of amyloid-β plaque and tau tangles in their neurons. Yet, these individuals retained the neurons, synaptic elements, axon geometry and cortical thickness needed for normal memory function. Brains of these individuals are referred to in the remainder of this chapter as *Alzheimer's-like brains*.

Observation of Alzheimer's-like brain with normal cognitive function is helping investigators re-evaluate factors involved in the Alzheimer's disease progression. The most dramatic non-neuron characteristic of Alzheimer's-like brain is a lack of reactive astrocytes. The large number of reactive astrocytes observed in the majority of Alzheimer's brains at postmortem examination is absent in Alzheimer's-like brains.

Other differences exist between Alzheimer's brain and Alzheimer's-like brain. Brains from Alzheimer's patients contain more of the structured beta sheet forms of amyloid-β and have amyloid-β plaques that are larger than those of Alzheimer's-like brains. Alzheimer's-like brains have less soluble tau in fluids surrounding synapses than Alzheimer's disease and other tauopathies.

Two neuron synaptic proteins, reduced about 50% in Alzheimer's disease compared to normal brain, remain unchanged in Alzheimer's-like brain. One of these proteins is a common component of the vesicles that store and release neurotransmitter, and the other is a component of the region of the synapse that contains neurotransmitter response elements.

Similarities between Alzheimer's brains and Alzheimer's-like brains include extensive tau tangles within neurons, a substantial quantity of amyloid-β plaque and equal amounts of single-strand flexible amyloid-β within synapses.

New Avenues for Progress

Revelation of the Nun Study and the Religious Orders Study that brains with intact reasoning ability sometimes display the classical markers of Alzheimer's disease is an important advance in knowledge about this illness. These brain control studies provide compelling evidence that amyloid-β plaque around, and tau tangles within, neurons are not sufficient to cause the massive loss of neurons associated with Alzheimer's dementia.

Some theories suggest rogue reactive astrocytes may kill rather than protect the neurons. Alternately, a mere switch of astrocyte phenotype to the reactive form without an increase in their total population may simply deprive neurons of the metabolic support they require for survival. An important distinguishing feature of Alzheimer-like brain was a lack of reactive astrocytes. Because astrocytes normally become reactive only in the presence trauma or a signal from activated microglia, it is important to discover the activation signal(s) astrocytes receive from microglia with progression of Alzheimer's disease.

In Alzheimer's-like brain astrocytes supportive of neuron well-being were preserved. Microglia evidently did not signal those astrocytes to participate in destruction of neurons, because the neuron population survived. Could it be that surveying microglia of Alzheimer's-like brain failed to receive the 'kill me' signal from neurons? Even though more needs to be learned about indicators of neuron distress at synapses, microglia may provide an excellent drug target right now to slow progression of Alzheimer's disease.

Microglia initiated inflammation is caused by secretion of molecules similar to those used by peripheral macrophage. Inflammation is usually a tightly controlled process where pro-inflammatory and anti-inflammatory molecules are secreted in a pattern that destroys an invader and then heals the tissue. When the system becomes unbalanced and pro-inflammatory molecules persistently dominate, tissue destruction occurs.

Until very recently peripheral macrophage was thought to be different than microglia because worn out cells are replaced by bone marrow stem cells. Microglia in contrast seeds neural tissue during embryonic development and the original mature cells provide expansions of the population for an entire lifetime. This odd capability of microglia, a mature cell, to renew itself forced scientists to re-evaluate their explanation of stem cells. Dogma was that only stem cells are capable of self-renewal. However, it is now known that several tissues outside the brain also host a fully differentiated macrophage population that is able to self-renew like microglia.

Self-renewing macrophage of tissues outside the brain may provide an accessible model for further investigation of microglia inflammatory response. Anti-inflammatory drugs specifically targeted to microglia present practical advantages for treating and delaying onset of Alzheimer's disease. A large data base is already available for extended use of various anti-inflammatory medications to aid in design of the research.

Drugs targeting microglia-induced inflammation are among those currently in pre-clinical and early phase clinical testing. It is very likely that the key to delaying onset and progression of Alzheimer's disease will be found in the way neurons and non-neuron brain cells connect, share and disengage. The old belief that a person only uses 10% of his/her brain cells because a mere 10% of brain cells are neurons has been proven false.

SUMMARY CHAPTER 8

- The classical theory of the last 30 years that proposes deposits of amyloid-β begin the sequence of events leading to death of neurons associated with Alzheimer's disease is being reevaluated

- The degree of intellectual impairment observed with Alzheimer's disease correlates with the number of neurons with tau tangles and decreased size of affected brain regions
- Sequential, progressive change in amyloid-β deposits and accumulation of tau tangles proceeds for an average of 10 years before mental impairment is experienced in patients with Alzheimer's disease
- Accumulation of amyloid-β into dense plaque formations throughout Alzheimer's brain is believed to be an outcome of decreased clearance of the normal brain molecule
- Abnormal tau in cerebrospinal fluid is not specific to Alzheimer's disease because an elevated quantity of tau is present in other forms of neurodegenerative diseases as well
- Loss of neurons in particular brain areas is a sequential process in Alzheimer's disease
- The first brain area affected by Alzheimer's disease is a region in the temporal lobes adjacent to the entorhinal cortex, followed in order by the entorhinal cortex, hippocampus, amygdala, areas of the temporal and parietal lobes associated with language function and then the prefrontal cortex

- Earliest changes observed in the shape and volume of the entorhinal cortex appear as soon as 8 to 10 years prior to memory loss symptoms
- The hippocampus shows atrophy 2 to 4 years before symptoms
- In people at high risk for Alzheimer's disease, a decrease in glucose metabolism can be detected decades before the onset of symptoms
- Brains of some healthy people retain intact neuron pathways yet display the markers of Alzheimer's disease, amyloid-β deposits and tau tangles
- The most striking difference between Alzheimer's-like brain and Alzheimer's disease brain is the portion of the astrocyte population that is in the reactive form
- Most astrocytes in late stage Alzheimer's disease are of the damage-control reactive phenotype rather than the functional-support protoplasmic and fibrous phenotype
- The large population of reactive astrocytes in Alzheimer's disease may play a role in killing neurons
- Alternatively, a switch in phenotype of astrocytes to their reactive form may deprive neurons of the support they require at synapses for survival
- Analysts report of the 583 drug candidates for Alzheimer's disease under development in 2015

by the pharmaceutical industry, the majority are still in the discovery and pre-clinical testing phase

- New drug candidates for Alzheimer's disease focus upon the latest evidence of how neurons and non-neuron brain cells interact with each other

Further Reading

THE FOLLOWING PAPERS ARE A FEW of the references consulted in writing this book. The list is far from exhaustive, but it provides a sample of the current scientific literature discussed. Most of these papers are available for free at http://www.ncbi.nlm.nih.gov/pubmed/. There is a search box at the top of the page at PubMed where you may type in either the PubMed Identification number (PMID) or the PubMed Central Identification number (PMCID) listed with each citation. Once you reach the free paper it can be downloaded in various formats.

2104 *Alzheimer's Disease Facts & Figures,* http://www.alz.org/downloads/facts_figures_2014.pdf, Alzheimer's Association Report

Bennett DA et al., *Overview and Findings from the Religious Orders Study,* 2012, Curr Alzheimer Res, 9:628-645, PMID: 22471860 PMCID: PMC3409291

Bergelson E and Swingley D, *The acquisition of abstract words by young infants,* 2013, Cognition 127:391-397, PMID: 23542412, PMCID: PMC3633664

Birren S.J. and Marder E., *Plasticity in the Neurotransmitter Repertoire,* 2013, Science 340:436-437, PMID: 23620040

Cakir T et al., *Reconstruction and flux analysis of coupling between metabolic pathways of astrocytes and neurons: application to cerebral hypoxia,* 2017, Theor Biol Med Model 4:48 Review, PMID: 18070347, PMCID: PMC2246127

Catterall WA, *Voltage-gated sodium channels at 60: structure, function and pathophysiology*, 2012, J Physiol 590:2577–2589, PMID: 22473783, PMCID: PMC3424717

Chan MY et al., *Decreased segregation of brain systems across the healthy adult lifespan*, 2014, Proc Natl Acad Sci USA 111:E4997-5006, PMID: 25368199 PMCID: PMC4246293

Corkin S., *Permanent Present Tense*, 2013, Basic Books, New York. ISBN 9780465031597

Dulcis D and Spitzer NC, *Reserve pool neuron transmitter respecification: Novel neuroplasticity*, 2012, Dev. Neurobiol. Vol. 72:1-15, PMID: 21595049, PMCID: PMC3192250

Eriksson PS et al., *Neurogenesis in the adult human hippocampus*, 1998, Nature Medicine Vol. 4:1313-1317, PMID 9809557, Free at Nature Medicine

Fowler PW et al., *Detailed examination of a single conduction event in a potassium channel*, 2013, J Phys Chem Lett 4:3104–109, PMID: 24143269, PMCID: PMC3797101

Greer PL and Greenberg ME, *From synapse to nucleus: Calcium-dependent gene transcription in the control of synapse development and function*, 2008, Neuron 59:846–860, PMID: 18817726, Cell Press Open Access

Häusser M et al., *Diversity and dynamics of dendritic signaling*, 2000, Science 290:739–744, PMID: 11052929

Hillman EM, *Optical brain imaging in vivo: techniques and applications from animal to man*, 2007, J Biomedical Optics, 12:051402 Review, PMID 17994863, PMCID PMC2435254

Kawakami R et al., *Visualizing hippocampal neurons with in vivo two-photon microscopy using 1030 nm picosecond pulse laser,*

2013, Scientific Reports 3: 1014, PMID: 23350026, PMCID: PMC3553458

Kuhl PK, *Early language learning and literacy: neuroscience implications for education*, 2011, Mind Brain Educ, Vol. 5, 128-142, PMID: 21892359, PMCID: PMC3164118

Lazarczyk MJ et al., *Preclinical Alzheimer disease: identification of cases at risk among cognitively intact older individuals*, 2012, BMC Medicine 10:127-139, PMID: 23098093, PMCID: PMC3523068

Lloren-Martin' M et al., *Selective alterations of neurons and circuits related to early memory loss in Alzheimer's disease*, 2014, Frontiers in Neuroanatomy 8:38-49, PMID: 24904307, PMCID: PMC4034155

McKenzie IA et al., *Motor skill learning requires active central myelination*, 2014, Science 346:318, PMID: 25324381

Medina M and Avila J, *The role of extracellular Tau in the spreading of neurofibrillary pathology*, 2014, Frontiers in Neuroscience, 8:113-119, PMID: 24795568, PMCID: PMC4005959

Merin-Serrais P et al., *The influence of phospho-T on dendritic spines of cortical pyramidal neurons in patients with Alzheimer's disease*, 2013, Brain 136:1913-1928, PMID: 23715095, PMCID: PMC3673457

Miller JF et al., *Neural activity in human hippocampal formation reveals the spatial context of retrieved memories*, 2013, Science Vol. 342, 1111-1114, PMID: 24288336 Free at www.sciencemag.org

Miller G, *Mysteries of the brain. How are memories retrieved?* 2012, Science 338:30-31, PMID: 23042864

Neher JJ et al., *Primary phagocytosis of neurons by inflamed microglia: potential roles in neurodegeneration,* 2012, Frontiers in Pharmacology 3:1-9, PMID: 22403545, PMCID: PMC 3288722

Panza F et al., *Is there still any hope for amyloid-bases immunotherapy for Alzheimer's disease?* 2014, Curr Opin Psychiatry, 27:128-137, PMID: 24445401

Perez-Nievas BG et al., *Dissecting phenotypic traits linked to human resilience to Alzheimer's pathology.* 2013, Brain 136:2510-2526, PMID: 23824488, PMCID: PMC3722351

Ranasinghe KG et al., *Regional functional connectivity predicts distinct impairments in Alzheimer's disease spectrum,* 2014, NeuroImage:Clinical 5:385-395, PMID: 25180158, PMCID: PMC4145532

Reece M, *Physiology: Custom designed chemistry,* 2012, Reece Biomedical Consulting LLC, Create Space ISBN 9781482326611

Salloway S et al., *Two phase 3 trials of Bapineuzumab in mild-to-moderate Alzheimer's Disease,* 2014, New England Journal of Medicine, 370:322-333, PMID: 24450891, PMCID: PMC4159618

Serrano-Pozo A et al., *A phenotypic change but not proliferation underlies glial response in Alzheimer's disease,* 2013, American Journal of Pathology 182:2332-2344, PMID: 23602650, PMCID: PMC3668030

Sertbaş M et al., *Systemic analysis of transcription-level effects of neurodegenerative diseases on human brain metabolism by a newly reconstructed brain-specific metabolic network,* 2014,

FEBS Open Bio 4:542-553, PMID: 25061554, PMCID: PMC4104795

Sieweke MH and Allen JE, *Beyond stem cells: self-renewal of differentiated macrophages*, 2013, Science, 342 342(6161):1242974. doi: 10.1126/science.1242974, PMID: 24264994

Snowdon DA, *Aging and Alzheimer's disease: Lessons from the Nun Study*, 1997, The Gerontologist, 37:150-156, http://gerontologist.oxfordjournals.org/content/37/2/150.long, PMID: 9127971

Spalding KL et al., *Dynamics of hippocampal neurogenesis in adult humans*, 2013 Cell 153:1219-1227, PMID 23746839, Free at CellPress Open Access

Swingley D, *The roots of early vocabulary in infants' learning from speech*, Curr Dir Psychol Sci 17:308-311, PMID: 20523916, PMCID: PMC2879636

Tomassy GS et al., *How big is the myelinating orchestra? Cellular diversity within oligodendrocytes lineage: facts and hypotheses*, 2014, Front Cell Neurosci Vol 8, Article 201, PMID: 25120430, PMCID: PMC4112809

Younes L et al., *Inferring change point times of medial temporal lobe morphometric change in preclinical Alzheimer's disease*, 2014, NeuroImage: Clinical 5: 178-187, PMID: 25101236, PMCID: PMC4110355

Wong AD et al. *The blood-brain barrier: An engineering perspective*, 2013, Frontiers in Neuroengineering, 6:1–22, PMID: 24009582, PMCID: PMC3757302

Glossary

Action potential – a pattern of voltage transients across an axon's membrane

Activated microglia – a form of microglia that destroys pathogens and removes dead cells

Adherens junction – a complex of proteins that tie the membranes of two cells tight together

Allosteric binding site – a position on a protein where the binding of a drug molecule forces a change in the shape of the protein

Alzheimer's disease – a progressive form of dementia characterized by abundant loss of neurons in brain areas associated with memory formation

Amyloid-β – a small peptide with an undefined function produced on the outside surface of brain neurons that accumulates as large insoluble deposits in the brains of Alzheimer's patients

Anterograde – direction of movement within neurons of material travelling away from the cell body

Apolipoprotein E – a brain protein that surrounds cholesterol as it is transported from one area to another

Arachnoid membrane – a membrane covering of the brain and spinal cord whose delicate fibers give it a spider web-like appearance

Astrocyte – a star shaped brain cell that works in partnership with neurons

Autobiographical memory – a form of memory similar to episodic memory but limited to events in a person's own life history

Axon – a long projection of a neuron's cell membrane that conducts chemical based electricity

Axon collateral – a branch off of an axon that travels through the brain in a different direction than the main axon

Axon hillock – the region of a neuron body that connects to the axon

Axon initial segment – the part of the axon closest to the neuron body that contains the axon's first set of voltage-sensitive sodium channels

Axon terminal – the far end of the axon away from the neuron body where neurotransmitter is stored

Blood brain barrier – blood capillaries that regulate transfer of molecules into and out of the brain

Brain inflammation – a condition in the brain produced by prolonged malfunction of a normal protective mechanism to destroy bacteria and virus

Brain stem – the part of the brain continuous with the spinal cord

Brodmann Areas – a map of cortical neuron patterns based upon marked differences in neuron configuration and connectivity

Carotid arteries – arteries of the neck that supply blood to the face and to the front and middle portion of the brain

Cerebellum – structure at the back of the brain critical for refinement of motor movements

Cerebral cortex – the six layers of large, intricately connected neurons that cover the brain hemispheres

Cerebrospinal fluid – fluid circulating through and around the brain and spinal cord that is produced by the choroid plexuses and the ependymal cells lining the surface of the ventricles

Choroid plexus – a tissue located in each of the brain's four ventricles that extracts nutrients from specialized blood capillaries and secretes a solution called cerebrospinal fluid

Circle of Willis – a safety net of blood vessels at the base of the brain where the anterior carotid and posterior vertebral circulations are connected by the posterior communicating arteries

Connectome – a proposed map of all the neuron connections in the human brain

Corpus callosum – a bridge of neuron axons that connect corresponding areas of the brain's right and left hemispheres

Cytoplasm – the fluid compartment within cells that is outside the nucleus

Cytoskeleton – the internal framework of a cell composed primarily of actin filaments and microtubules

Dementia – deterioration of a person's intellectual abilities

Dendrite – a branched extension of a neuron body that collects and conducts impulses from adjacent neurons inward toward the cell body

Dendritic spikes – membrane voltage transients originating in neuron dendrites similar to action potentials

Dendritic spines – thorn-like projections of dendrite membrane where synapses are located

Diencephalon – the division of the brain that includes the thalamus and hypothalamus

Diffusion – in chemistry the relocation of molecules within a solution away from an area where they are in high concentration

Dura mater – a double layer of thick fibrous membrane between the brain and the bone of the skull

Electroencephalogram, EEG – the difference in field potentials detected by pairs of scalp electrodes

Ensemble coding – action potential coding of information carried out by populations of neurons working together

Entorhinal cortex – an area of the temporal lobe of the brain near the hippocampus that acts as an interface between the hippocampus and the neo-cortex where memory is stored

Ependymal cells – small cuboidal ciliated cells lining the surface of brain ventricles that secrete and absorb cerebrospinal fluid

Episodic memory – the memory form that recalls specific events, people, situations and personal experiences

Equilibrium potential for potassium – a calculated transmembrane potential for cells with passive K^+ channels but no passive Na^+ channels

Equilibrium potential for sodium – a calculated transmembrane potential for cells with passive Na^+ channels but no passive K^+ channels.

Event-related Potentials, ERPs – electrical activity measured by pairs of scalp electrodes that is time locked to presentation of a sensory stimulus such as a picture

Excitatory neuron – a neuron that releases a neurotransmitter that causes dendritic spikes in postsynaptic neurons

Explicit/declarative memory – memory that can be recalled at will

Fibrous astrocytes – glial cells that position themselves throughout all white matter where they make contact with axons at Nodes of Ranvier

Foramen magnum – a large hole in the base of the skull where the spinal cord connects with the brain stem

Functional Magnetic Resonance Imaging, fMRI – a variation of MRI that measures flow of oxygenated blood into small volumes of brain tissue

Functional Near Infrared Spectroscopy, fNIRS – an imaging method that measures changes in brain blood flow by recording near infrared thermal radiation penetrating through brain tissue, bone and skin

Gap junctions – membrane structures shared by two cells with *a central open channel* that connects their cytoplasm

Glycogen – an storage form of glucose within cells

Gray matter – brain tissue dominated by large clusters of neuron cell bodies

Gross anatomy – the external features of a dissected tissue or organ

Hippocampus – the ancient part of the cerebral cortex that receives multiple inputs from sensory organs and uses that information to code new memories

Implicit/procedural memory – memory below the level of conscious awareness that forms as an experience is repeated; recognition of the category to which a piece of music belongs is an example of implicit memory

Inhibitory neuron – a neuron that releases a neurotransmitter that blocks dendritic spiking at a postsynaptic neuron

Interneuron – a small inhibitory neuron that regulates the activity of larger neurons

Interstitial fluid – the protein free fluid that surrounds cells

Ion – an atom that lacks a match in its number of positive particles, protons, and negative particles, electrons

Ion exchange pump – an energy consuming protein that moves ions across a cell membrane against their concentration gradient

Ligand – a small molecules who's binding to cell proteins regulates cell performance through a variety of mechanisms

Local field potential – an electrical potential in the brain created by opening of membrane ion channels at multiple synapses and the resultant movement of ions within the tissue

Long term memory – includes both explicit/declarative and implicit/procedural memory

Long term potentiation – an increase in the ability of brain synapses to respond to neurotransmitter after receiving a rapid burst of signaling activity induced by external electrodes

Macrophage – a cell of the body's immune system that surrounds pathogens and dead cells and recycles their components

Magnetic Resonance Imaging, MRI – a noninvasive procedure that produces an anatomic picture of large brains structures in virtual slices of living tissue

Magnetoencephalography, MEG – a noninvasive procedure that records magnetic fields produced by electrical currents flowing through brain circuits

Medulla oblongata – another name for the axons making up the brain stem the means long white rope

Meninges – protective membranes covering the entire brain and spinal cord

Mesencephalon – a division of the brain deep in the center of the organ also called the midbrain that coordinates complex reflex reactions

Metencephalon – the cerebellum and the pons

Microglia – non neuron cells of the brain that monitors neuron well-being and provide the brain's immune response to pathogens

Mitochondria – structures in the cytoplasm of cells containing enzymes necessary for conversion of food to useable energy

Myelencephalon – another name for the brain stem

Myelin – the white fatty material covering neuron axons

Neo-cortex – through evolution the most recently developed part of the cerebral cortex

Nerve – a bundle of neuron axons

Neural stem cell – partially differentiated brain stem cell capable of providing replacement neurons and oligodendrocytes

Neurogenesis – the birth of new neurons from neural stem cells

Neuroglia – the oligodendrocytes and astrocytes of the brain

Neuron – an individual electrical cell of the brain or spinal cord

Neuron circuit – an organized group of neurons that operate together as a single unit

Neuroplasticity – the brain's ability to rearrange its dendrites and dendritic spines in response to sensory stimulation such as sound and light

Neurotransmitter – a chemical released by an axon terminal to signal to another cell

Nodes of Ranvier – bare patches of axon between myelin layers

Nucleus – in brain a cluster of neuron cells bodies in the white matter beneath the cerebral cortex; within cells a compartment that houses genetic material

Oligodendrocytes – brain cells that wrap neuron axons with myelin

Passive ion channel – a cell membrane pore for ions that remains constantly open

Phonemes – the smallest units of speech in a language capable of conveying a distinction in meaning

Pia mater – innermost of the membranes covering the brain adhering to the outer surface of the cerebral cortex and forming a sheath around arteries entering the brain

Pons – a span of brain tissue that connects the cerebellum to the cerebral hemispheres

Positron emission tomography, PET – a brain imaging method that requires injection of a radioactive molecule that can cross the blood brain barrier and emit a signal after binding to a brain substance

Postsynaptic density – a thickening of the neuron membrane that contains neurotransmitter response elements

Prosencephalon – the forebrain in the embryo that matures into the retina, optic nerve, iris, cerebral hemispheres, thalamus and hypothalamus

Protoplasmic astrocytes – glial cells of the gray matter that surround neuron cell bodies

Radial cells – embryonic stem cells that produce neurons, astrocytes and oligodendrocytes

Reactive astrocytes – astrocytes that respond to damage in the brain by removing excess toxic glutamate, producing anti-oxidants and walling off damaged tissue from healthy tissue with scar formations

Receptor – a generic term for a broad class of proteins activated by specific chemicals called ligands that regulate cell performance through a variety of mechanisms

Retrograde – direction of movement within neurons of material travelling toward the cell body

Rhombencephalon – in the embryo the last brain division which matures into the cerebellum, pons and brain stem

Semantic memory – memory of facts like number of hours in a day and meaning of a word

Short term memory – another name for working memory which is a dynamic form of memory combining many things from past learning together with present experiences

Superior sagittal sinus – an area between the layers of dura mater where venous blood and cerebrospinal fluid pool on their way back to the heart

Surveying microglia – the form of microglia that monitors neuron synapses and provides repair to damaged neurons

Symporter – a protein pump that moves two or more different molecules, or ions, in the same direction across a membrane where at least on molecule is moving down its concentration gradient and one is being moved against its concentration gradient

Synapse – structure at the place where neurons contact each other or contact other cells

Synaptic cleft – the gap at a neuron synapse of about 20 nanometers

Tau – a soluble protein within neurons essential for regulation of the microtubules that move material from place to place in the cell

Tau tangles – a form of tau that precipitates within neurons as pairs of helical filaments and aggregates into a structure visible with a light microscope

Telencephalon – in the embryo the far end of the neural tube that matures into the right and left hemispheres

Thalamus – the part of the brain that serves as an entry point for information coming to the brain from the rest of body

Tract tracing – charting the path of neuron axons through brain tissues by injecting dye and observing its movement within the cell

Transmembrane potential – the difference in electrical field potential created by ions and charged proteins on two sides of a cell membrane

Ventricles – the four hollow chambers in the center of the brain

Vertebral arteries – branches of the large arteries supplying blood to the shoulders, lateral chest and arms that run through the cervical vertebrae and into the head where they perfuse the back of the brain

Virchow-Robin Space – the area around brain arteries created by the pia mater sheath

Voltage-sensitive ion channel – a channel for ions through a cell membrane that opens and closes in response to changes in the transmembrane potential

White matter – areas of the brain with few neuron cell bodies but many neuron axons

Working memory – a dynamic form of memory combining many things from past learning together with present experiences

ABOUT THE AUTHOR

Margaret Thompson Reece PhD, former Senior Scientist in academic medicine and Chief Scientific Officer at Serometrix LLC, heads Reece Biomedical Consulting LLC.

Dr. Reece helps students who struggle to figure out how to study human anatomy and physiology through her website http://www.medicalsciencenavigator.com/media-kit/, and her speaking and writing. Dr. Reece consults privately with a small selected group of students. She lives in upstate New York.